古早中国锁具之美

颜鸿森 著

海南出版社
· 海口 ·

古早中国锁具之美
颜鸿森著
中文简体字版© 2019年，由海南出版社有限公司出版。

版权合同登记号：图字：30-2017-079号
图书在版编目（CIP）数据

古早中国锁具之美 / 颜鸿森著 . -- 海口 : 海南出版社 , 2019.11
ISBN 978-7-5443-8921-1

Ⅰ. ①古… Ⅱ. ①颜… Ⅲ. ①锁具 – 文化史 – 研究 – 中国 – 古代 Ⅳ. ① TS914.211-092

中国版本图书馆 CIP 数据核字 (2019) 第228322号

古早中国锁具之美
GUZAO ZHONGGUO SUOJUZHIMEI

作　　者：颜鸿森
监　　制：冉子健
责任编辑：张　雪
策划编辑：冉子健
执行编辑：邓博文
装帧设计：衣　波
责任印制：杨　程
印刷装订：天津联城印刷有限公司
读者服务：武　铠
出版发行：海南出版社
总社地址：海口市金盘开发区建设三横路2号 邮编：570216
北京地址：北京市朝阳区黄厂路3号院7号楼102室
电　　话：0898-66830929　010-87336670
电子邮箱：hnbook@263.net
经　　销：全国新华书店经销
出版日期：2019年11月第1版　2019年11月第1次印刷
开　　本：787mm×1092mm　1/16
印　　张：8.5
字　　数：100千
书　　号：ISBN 978-7-5443-8921-1
定　　价：49.90元

序一

在警匪剧中最惊心动魄的情结莫过于开锁，锁住的是一个秘密，解锁的是一个间谍，锁的构造代表着机巧，解锁的方法代表着智慧。斗智斗勇的瞬间是任何文学作品和影视剧里的高潮。

我是从事古代家具研究的，家具与锁的关系是密不可分的，但是家具用锁绝对不是为了防盗。因为中国人最亲密的关系就是一家人，从来不会把家人当贼。那么锁在家庭中有什么意义呢？

首先是美观，中国的明清家具用料讲究紫檀花梨，做工讲究榫卯结构，最注重的是外型漂亮，这就离不开画龙点睛的锁。在家具的专业领域中，都用“型、艺、材、韵”作为评价标准，其中排在第一位的“型”就是家具的外形漂亮不漂亮，这其中就包含了铜活，铜活中最出彩的就是一把铜锁。

其次是权威，家具上的锁只有家中最具权威的家长握有钥匙，当家中年终进行财产分配时，白胡子老爷爷会从腰中解开挂绳拿出钥匙打开柜子的锁，全家人的眼神都锁定在老爷爷颤颤巍巍的手里的那把钥匙之上。

中国古代的锁具都是铜匠单独打造，所谓一把钥匙开一把锁。所以对锁具的研究就变得非常困难，因为研究的基础是要收藏到大量的实物。非常高兴看到了颜鸿森先生的专著，为我们解开中国古代工匠的奇巧手艺提供了钥匙。

于鸿雁

古典家具宫廷烫蜡技艺非遗传承人

全国工商联文物艺术品商会副会长

序二

锁，区分出两个概念的世界："我的"和"不是你的"。"我的"是原始的个人体验，在那个世界里，凡是感官所及，都是我的；"不是你的"是超越感官的世界，建立在较高层次的反思历程上，成为一项防卫的机能。有了锁，人类的精神文明就不再只是"有"与"无"的关系而已。更重要的是不但要有，而且要把"有"储存，把"有"珍藏，要保护已经"有"的，要防范"有"被人偷走，所以为保护眼睛看不到的"有"，就设计一把锁，把看不见的"有"关起来，拥有那把钥匙的人，就是"有"的主人。而"不是你的"就表示你没有开锁的身份与资格，和持钥之主不在等同的社会地位上。人际关系变了，人类文明的形式也跟着变了。

锁住的岂止是有形的东西。那记在记事本上的抽象思维，那日记里喃喃的含情苦思和一片片枯黄枫叶的相思情怀，加上一些不足为外人道的零碎事物，容我加一把锁，锁住个人的隐私。

啊，隐私！这是个人化社会文明中的精神表征。隐私是一种无形的灵魂，是不可被侵犯的，是超越亲情的。所以，锁在里面的，是个人的秘密档案、无限遐思、无尽哀怨，曾有忧伤，也有快乐的片段，锁在心灵深处，不许外人(就是亲生父母也不例外)窥探。

如何锁住这些有形无形的"个人所有"呢？大锁、小锁、方锁、圆锁、虾尾锁、鱼形锁、胡琴锁、龙蛇锁、飞鸟锁，造型各异，且细雕巧铸，琳琅满目，锁住人类生活的多样性表现。再看锁内的机关装置，有的簧片往上，有的打横，有的匙沟还长出小牙，以防外人无钥自通。这种种的设计，更展示了人类心思灵巧，智慧已由计算演进到算计的更高层次了。

收藏锁的人，也因而神秘了吗？一点儿也不！

颜鸿森教授一生俭朴，个性沉稳，坚守专业，又勇于任事，让人有一夫当关、万夫莫入的感觉。他本身就像是一把坚固不易破解的锁，所以他收藏锁，理所当然。颜教授专攻机械原理，对古今中外的机械构造，都了如指掌。他平日木讷寡言，一看到各式机件，就会眼睛发亮。锁的内在机件如何组合，外形又如何与功能搭配，他一介绍起来，就口若悬河，滔滔不绝，尤其谈到他追求锁的经历，更是引人入胜。唯有他，才能讲得这么精彩。

我有一次问颜教授，有形的锁锁得住无形的现实世界吗？那锁住一室春光而不令其外泄的，会是怎么样的一把锁呢？为之序。

曾志朗

中国台湾研究院副院长

2003 年 3 月 19 日于台北

PREFACE

Locks, which are utilized to protect something for its security, conceptualize the world into two ideas: "mine" and "not yours". "Mine" is the primitive personal experience. In that world, all feelings of sense organs are mine. "Not yours" is a world that transcends the sense. It is built on a higher level of introspection and becomes a defense mechanism. The meaning of a lock is more than the "existence" and "on existence" in the civilization of mankind. The "existence" should be stored, treasured and protected, so that it will not be stolen. In order to protect the "existence" that is out of the owner's sight, the lock is then designed to lock the "existence". The one who owns the key is the master of this "existence". "Not yours" means that you do not have the identity and qualification to unlock it. You and the key owner are persons of different social status. Therefore, the personal relations are changed, so is the form of the whole civilization.

The locked existence is far more than the tangible things. The abstract thinking written in a notebook, the murmuring tenderness and love hidden in a diary, the withered maple leaves of lovesickness, and a fragmentary statement that cannot be described are all the things to be locked. For me, a lock is used to lock my personal secrets, my privacy.

Privacy, the spiritual symbol of the personalized civilization, is an invisible soul, cannot be invaded, and even transcends the blood relationship. What is locked inside is a personal secret file. There are unlimited wild and fanciful thoughts, sadness, sorrows, and happiness. They are all locked deeply in the soul and no one could pry about it, not

even one's parents.

How can we lock those tangible and intangible "personal belongings"? Many locks with different shapes can be used: big locks, small locks, square locks, and round locks. There are locks that look like a shrimp tail, a fish, a two-stringed bowed instrument, a dragon/snake, and a flying bird. They are all sculptured exquisitely and delicate. Its beauty is dazzling. It also locks the diversity of our lives. Look inside its apparatus. There are the splitting-springs equipped upwards and crosswise. Some have little teeth on the key slot, so no one can unlock it without the right key. This kind of design shows the dexterity of mankind. Wisdom is developed into a much higher level, from simple calculation to a pre-determined calculated scheme.

Does the one who collects locks also become mysterious? No, not at all!

Prof. Hong-Sen Yan is very thrifty in his daily life and dedicated in his work. He always takes all challenges and can be fully trusted for all matters. He collects all locks because he, himself, is a solid lock that cannot easily be cracked and opened; in addition, he is specialized in the principle of machinery and understands all mechanical structures. Dr. Yan, the collector, is both natural and sensible. He is often quiet, but when he sees various mechanical devices, he becomes extremely excited. He can explain, all in details, the internal mechanisms of the locks and shows how to assemble it or describes the external pattern of the lock and its function. He cannot stop explaining while he introduces the locks, mostly when he is talking about researching the locks. And, only he can make the explanation so interesting.

I once asked Prof. Yan, "Could the tangible locks keep the intangible real world locked? And, what would a lock be, if it could block the secrets inside the room and would not let it blab out?" A preface for his book describes the intriguing locks.

Dr. Ovid J. L. Tzeng

Vice-President, Academia Sinica

19 March, 2003, in Taipei, Taiwan,China

编者序

“锁”，这个汉字代表的物件曾在几千年内保护着中国古人们。在那些天灾人祸接踵而至的年代，是锁隔绝了咆哮的风雨与野兽，为人们开辟出一片属于自己的空间，守住尘世中的一片安隅。

然而，锁仅仅是人们的一片净土吗?

不。中国人把盛世形容为“路无拾遗，夜不闭户”。

但又有一说：“锁防君子，不防小人。”

在中国人的记忆里，锁是人们对于秩序正常运行的一种认可，但也是防范；锁是人们对所有权的一种心照不宣，也是对主人的尊重；锁是中国自然经济里的血脉烙印，也是关于乡亲邻里的地域情缘；锁更是夜幕降临、家人齐聚时燃起的袅袅炊烟，是月上枝头、人声初静后的寒灯数点……

若是抛却帝王家歌功颂德的名篇，踏入那抛却纷扰的桃源世界，人们对于时代的回忆就全都寄于区区一锁之上。

锁，是我们民族的记忆。

二十世纪四五十年代，西方栓销制栓锁陆续进入市场。在那个物资紧缺的年代，新式锁因为价格便宜、保密性强，很快风靡全国。传统的中华古锁终于在新时代完成了自己固有的使命，随着众多古老的记忆一起，逐步退出了历史舞台。

不过区区数十年，便像是争相印证马斯洛关于需求的至理名言一般，人们一旦不再囿于温饱，便开始追寻其存在的缘由与历史，在那些陈旧的古董中翻箱倒柜，试图寻找他们的时代存在过的证据。在发掘的过程中，他们又逐渐被博大精深的中国传统文化所吸引。

这便是《古早中国锁具之美》想向您传达的故事，它将带领您回到那个草木含情的时代，回味华夏先民的民俗风韵。

据出土文物考证和历史文献记载，中国锁具发展至今已有近5000年历史。可以说，在我们的文明诞生伊始，锁具就应用于先民的生活中，帮助先民抵御着洪荒时代难以预料的自然威胁。

随着社会形式的变迁和科技进步，锁的形态与功能日新月异。从抵御自然，到阻止人类；从封闭居所，到锁柜、锁箱；从设计古朴的木锁，到形态丰富的花旗锁……这仅仅方寸之物，却被赋予了七十二变之能。然而，古锁仅代表着形而下的物质世界吗?

元稹曾咏菊："不是花中偏爱菊，此花开尽更无花。"对古人而言，菊不仅仅是一种花，更是一种傲立寒秋的忠贞与勇气。

锁又岂能是区区一锁呢?

在此书中您将看到，锁的一雕一琢、一方一角，都包含着先民们的思想精粹与手工艺者们的无上匠心。

锁的外形标志着它的寓意。《芝田录》记载："门钥必以鱼者，取其不瞑目，守夜之义。"鱼亦作年年有余之吉利。同时，大量古锁镌刻有文字或者图案，如：状元及第、长命富贵、麒麟送子、龙凤呈祥等。古锁又寓意财富，是以有"驴驮钥匙马背锁"之说。后衍生至爱情永恒，家族永续，身体永康……在古代工艺者的精湛雕工下，古锁是百姓们朴素愿景的承载物，似乎一切良愿尽寓其中。

钥匙孔的形状又标志着社会阶级的分化。其中，"一""上""士""工"

“古”“尚”“吉”“喜”“寿”等字都是比较常见的钥匙孔形状。“一”字孔锁是庶民使用的锁具；“士”字锁是文人、士大夫常用的锁具；“吉”字锁则是达官贵人所用之物。至于帝王将相之锁，则是平民永不能跨越的无边沟堑。

它们还长存于诗人们的妙笔之下，刘禹锡的“新妆宜面下朱楼，深锁春光一院愁”；唐寅的“深院青春空白锁，平原红日又西斜”；黄庭坚的“巫峡高唐，锁楚宫朱翠”……锁又成了诗人们欲语还休的美妙意象，锁住了萦绕心头的一江春水，还有对春色可望不可即的无限愁思……

它是中国人民对于太平盛世的殷切盼望，是华夏民族对于现存社会的深入诠释，更是一切美好事物的精神表征！而我们对于它的认识却无比浅薄。

对我们来说，古锁到底是精致小巧的，所以它才能在如今保有极高的收藏价值；但若说是小巧的，它腹内又有着如何之广阔的乾坤，能够承载中华文明这么多年的历史记忆与文化底蕴?

这本书将给你答案。

它将带你一起回溯那些白驹过隙般的历史片段，在时光的寸缕中回顾古早中国锁具的发展历程。在这个过程中，你会切身领略古早中国锁具之美，更会探索中国五千年来风雨兼程的文明之路。

自序

1978 年在美国普渡大学留学时，指导教授阿伦 · S. 霍尔博士曾带我参观他所收藏的秤，林林总总，不下百个，让我涌起收藏点儿什么的念头。虽然当时未能确定藏品的对象，但是 1980 年获得博士学位后，我构思出藏品要合乎下列五个要件：必须是个机构，以符合研究专长；必须和本土文化有关，以方便深入探讨；不能太大，否则在空间上难以配合；不能太贵，否则在财力上难以负担；不能太多，否则在特色上难以突出。

1986 年秋某日傍晚，在台北火车站附近的地摊上，我偶然间看到一个不起眼的古铜锁，顿时热血沸腾，立即决定以古早中国锁具为收藏的对象。从此以后，我每每利用闲暇，访遍国内外不少城乡的博物馆、古玩街、旧货摊、跳蚤市场，甚至大街小巷，努力寻觅古锁的踪迹，如今约有 700 件藏品。

我个人认为，收藏的真正价值不在于购买与保管物件，而是鉴赏与研究，并有所发现。因此，自 1990 年起，在公忙之余，我投入到了古锁的研究中，除了在各处博物馆参观古锁的展示外，也到各地图书馆寻找古锁的资料。有关古西洋锁具的展示与文献，虽然不多，但有其完整性。然而，有关古中国锁具的保存与展示，既稀少又残缺不全，而关于古中国锁具之历史发展与特征说明的文献与书籍，更是付之阙如。亦因如此，才激发了我的勇气，不自量力地执笔写书。

本书根据个人十多年来对锁具的收藏与研究经验，系统化地介绍中国锁具的发展与特征，内容包括绪言、历史发展、古锁语源、古锁类型、古锁外形、锁体雕花、锁体材质、古锁构造、簧片构形（构造与形状）、钥匙、钥匙孔、古锁开启等十二个单元，并配了一百多张精美的照片，以展示出古早中国锁具之美。由于锁具是属量产的工艺件，非具唯一性的艺术品，因此并无所谓的赝品，而本书有些锁具选用近年来的复制品，以便较完整地介绍古锁的类型与特性。

承蒙前台湾省文化处陈倬民处长及台湾省立台南社会教育馆林案倔馆长的支持，本书于 1999 年 6 月以中文首次出版。其后，不时有读者与友人建议发行英文版，以方便国际人士了解中国锁具，加以这三年多来亦搜集了不少精致的古锁，因此在百忙中排除万难，以中英文完成本书的再版。此外，此书之成，承张坚浚、查建中、邹慧君及其他师友的鼓励与贡献，中国台湾成功大学研究生黄馨慧与中国台湾科学工艺博物馆同仁李如菁的投入与协助，以及陈安纯与勾淑婉的英文翻译与帮忙，并此致谢。

期望本书的出版具抛砖引玉之效，引起大众对古锁的兴趣，并启发更多有关古早中国锁具的搜集、保存、展示、研究及发表，亦祈各界读者专家赐予指教。最后，谨以本书献给教育著者成长、但已仙逝的霍尔教授。

颜鸿森

中国台湾成功大学教授

中国台湾科学工艺博物馆馆长

2002 年 12 月 12 日于台南

AUTHOR'S PREFACE

When I studied at Purdue University in 1978, Professor Allen S. Hall Jr. showed me more than one hundred scales of various kinds that he collected. This inspired me to have a collection of my own. Although I had no idea what it might be, but when I completed my Ph. D. degree in 1980, I came up with a conclusion that it should meet the following five requirements: should be a mechanism, should be related to my native culture, should not be too large, should not be too expensive, and should not be too common.

One day at dusk in the fall of 1986, I happened to find a shabby, old brass lock at a vendor stall near Taipei Railway Station. That moment I was extremely excited and instantly decided to collect ancient Chinese locks. Ever since the day, I have visited quite a number of museums, antique shops, flea markets, second-hand merchandize stores, ... trying to find as many types of ancient Chinese locks as possible. Now I have around 700 pieces in my collection.

The true value of a collection lays in the appreciation, study and discoveries of the object, not merely through purchase and preservation. Therefore, I began my research on ancient Chinese locks in 1990. I visited many museums all over the world to see the exhibitions of ancient locks, and I buried myself in the piles of books in the libraries. I found that the introduction, exhibition and documentation of western locks, though not large in quantity, have the integrity and organization. However, information on the exhibitions and preservation of ancient Chinese locks are very little and incomplete, not to mention any academic document or text on the historical development or any introduction to the characteristics of ancient Chinese locks. These facts instilled enough courage in me to start working on this book.

This book introduces the development and characteristics of ancient Chinese locks systematically based on my collection and research in the past years. It includes the following twelve chapters: introduction, historical development, linguistic origin of locks, types of locks, shapes of locks, engraving of locks, materials of locks, mechanisms of locks, configurations of springs, keys, keyholes, and opening of locks, with more than one hundred elegant photos to support and demonstrate the beauty of ancient Chinese locks. Since locks are craft items with mass production, not unique art pieces, broadly speaking, there are no fake products. And, in order to illustrate the types and characteristics of ancient locks better, this book adopts some locks that were manufactured in recent years.

Here, I would like to thank the Cultural Division and the Tainan Social Education Hall of the ex-provincial Taiwan Government for their kind support in publishing the 1st edition of this book in Chinese in June 1999. In the past three years, many friends strongly suggested me to provide an English version to serve international readers. And along with many delicate new collections, I was finally able to manage my time to revise the book with both Chinese and English languages. On the way to complete this task, I owe sincere appreciation to the encouragement and contribution endowed on me from my friends: Paul Chang, John Cha, Hui-Jun Zou, Lily Chen, Karen Kou, and many others. Special thanks have to go to my graduate student Hsing-Hui Huang at Taiwan

Cheng Kung University (Tainan, Taiwan) and my colleague Ju-Ching Li at Taiwan Science and Technology Museum (Kaohsiung, Taiwan) for their endless assistance.

I truly hope the publication of this book will induce more interest from people toward ancient locks and I expect to see more and more presentations, collection, perseverance, exhibitions, and research in this field. I will be thankful to receive all responses and corrections from the readers.

Finally, I would like to dedicate this book to late Professor Allen S. Hall, Jr. who educated me and inspired me in so many ways.

Dr. Hong-Sen Yan

Hong-Sen Yan

Chair, Cheng Kung University

President, Science and Technology Museum

December 2002, at Tainan, Taiwan,China

目录

CONTENTS

绪言
INTRODUCTION
002

历史发展
HISTORICAL DEVELOPMENT
008

古锁语源
LINGUISTIC ORIGIN OF LOCKS
012

古锁类型
TYPES OF LOCKS
017

古锁外形
SHAPES OF LOCKS
021

锁体雕花
ENGRAVING OF LOCKS
039

锁体材质
MATERIALS OF LOCKS
064

古锁构造
MECHANISMS OF LOCKS
072

簧片构形
CONFIGURATIONS OF SPRINGS
078

钥匙
KEYS
084

钥匙孔
KEYHOLES
090

古锁开启
OPENING OF LOCKS
098

参考文献 REFERENCES
110

关键词 KEY WORDS
112

探索古中国锁具不为人知的美丽世界。

Explore the beaultiful world of unnoticed ancient Chinese locks.

绪言

INTRODUCTION

数千年来，锁具被广泛地使用在日常生活中，其发展始于人们在心理上对个人、群体或群体中之个人在安全上的实际需求。随着社会的变迁与科技的进步，锁具的功能愈加完备，不仅设计形式日新月异，且其制作精度也逐渐提高。此外，锁具应用范围，由往昔仅用于锁门、锁柜、锁箱，进而扩展到锁办公桌、锁保险柜、锁各型车辆等。然而，当人们在享受使用方便和安全的锁具时，往往不在意锁具的由来，更不知古锁的典雅优美，以及其所富含的艺术价值。

中国历史悠久，但是有关锁具的文献记载与实物保存，却相当缺乏。历代以来，锁具的制作，皆由地位卑微的锁匠为之，虽不乏巧手与奇品，但几乎都名不见经传。古锁是民俗文化的瑰宝，但由于其不易被注意到的性质，

Locks have been widely used in our daily life in the past thousands of years. The development of locks arises psychologically from practical needs on safety for individuals, for groups, or for individuals within groups. Following the progressive development of the society and the technology, the function of locks is getting complete. Not only the design of locks is greatly improved, but also the manufacturing of locks is more precise. Furthermore, the usage of locks has been extended from doors, chests, and boxes to desks, safes, vehicles, etc. However, when one is enjoying using the convenient and safe locks, he habitually neglects the development and history of locks. He usually pays no attention to the beauty of ancient locks. In addition, he normally ignores the artistic values of ancient locks.

Though with a long history, the related documents and material object of ancient Chinese locks are quite insufficient. Over the centuries and through dynasties, the manufacturing technology of this art-craft was strictly confined to locksmiths who were base in their social status; so even there were a good number of very delicate ones been created, the authors were mostly not

箱子与挂锁 chest and padlock

少有收藏家以古锁为寻觅的对象，亦鲜有学者以古锁为研究的主题。再者，当人们在打开古代的箱箧、橱柜或建筑物时，通常是将附在其上的锁具破坏，以获取内部的物品，未曾意识到这些毫不起眼的古锁具有保存与研究的价值，更不会注意到这些历经岁月沧桑之破旧物件所蕴藏的美丽与意义。在极少受到重视的情形下，现存古早中国锁具的数量逐渐减少，且散失的速度日益加快，而受到妥善保存的传

acknowledgeable because their names were never mentioned. Ancient Chinese locks are treasures of the Chinese civil culture, but for their hardly noticeable nature, very few curio collectors set their eyes on locks, and very few scholars focused their study on locks. When one was trying to open treasure boxes, closets, or buildings, he normally destroyed the locks to get the items inside. He had no sense that these shaggy locks have the value of preservation and research. He realized nothing about the beauty and historical meaning of these rusted metal pieces. In such a consequence, ancient Chinese locks have been drifting in the current of time unnoticed and the amount of ancient Chinese locks is getting fewer in a quicker speed. As a result, very few good ancient Chinese locks have been cared in good environment.

Ancient Chinese locks are mechanical padlocks, mostly key-operated bronze locks with splitting springs and partially keyless letter-combination locks. The major features of ancient locks are the types of locks, the shapes of locks, the engravings of locks, the materials of locks,

世古锁，更是屈指可数。

中国的古锁为机械式挂锁，大部分是以钥匙开启的簧片青铜锁，少部分是不需钥匙的文字组合锁。古锁的特征很多，主要包括：锁具的类型、锁具的外形、锁体的雕花、锁体的材质、锁具的构造等。而簧片锁更有以下几项重要的特点：弹簧片的构形、钥匙的数目与钥匙头的形状、钥匙孔的位置与形状、锁具的开启方式等。簧片构造锁大多用铜或铁做成锁体与锁栓，锁栓上有数片分离的弹簧，钥匙进入锁体后，能挤压钳制张开的弹簧片，使锁栓与锁体分离。

与西洋锁相比较，古早中国锁具最大的特点在于：有些锁具的钥匙不易

and the structure of locks. And, the character-istics of the splitting spring locks are the configuration of the splitting springs, the shape of key-heads and the number of keys, the location and shape of keyholes, and the opening of locks. Typical ancient Chinese locks are splitting spring locks opened by keys. The lock body and the sliding bolt are mostly made of bronze or iron and several splitting springs are attached to the sliding bolt. Once the key is inserted, the splitting springs are squeezed down and the sliding bolt could be released from the lock body.

Compared with western locks, the major charac-teristics of ancient Chinese locks are: some designs are not easy to insert the keys into the keyholes, and some designs are even not easy to find the keyholes.

This book systematically introduces the development and characteristics of ancient Chinese locks based on author's collection and

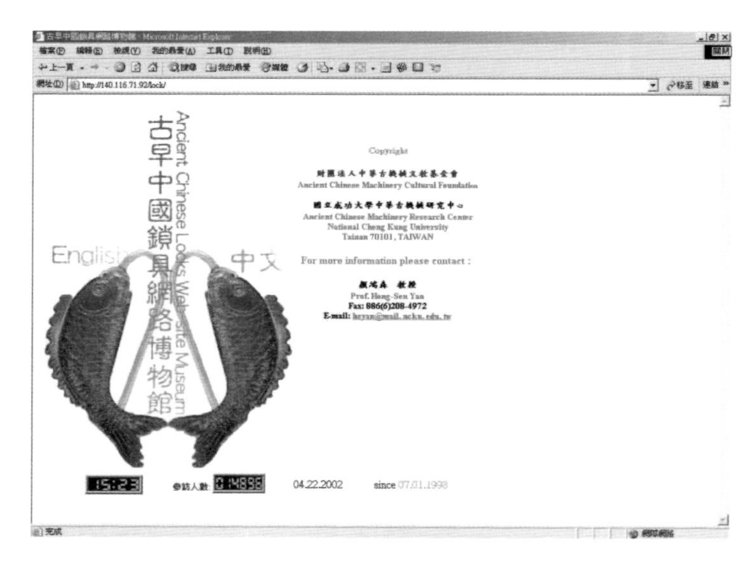

网络博物馆 web-site museum
http://www.acmcf.org.tw/chineselocks

明代铜锁 brass lock of Ming Dynasty

直接插入锁体上的钥匙孔，有些锁具的钥匙孔则隐藏在机关之下难以发现。

本书根据笔者十多年来对锁具的收藏与研究，以及近年所建立的“古早中国锁具网络博物馆”，系统地介绍了古锁的发展与特征，旨在尽一份微薄的心力，弥补一点有关锁具的历史文化空缺，并尝试让社会大众认识古早中国锁具的美丽。

research, together with the established "Web-site Museum of Ancient Chinese Locks", in the past years. The author wishes to share his duty to fill in the historical and cultural gaps regarding ancient Chinese locks, and most of all to provide the readers a glimpse at the beauty of ancient Chinese locks.

汉代铁锁 iron lock of Han Dynasty

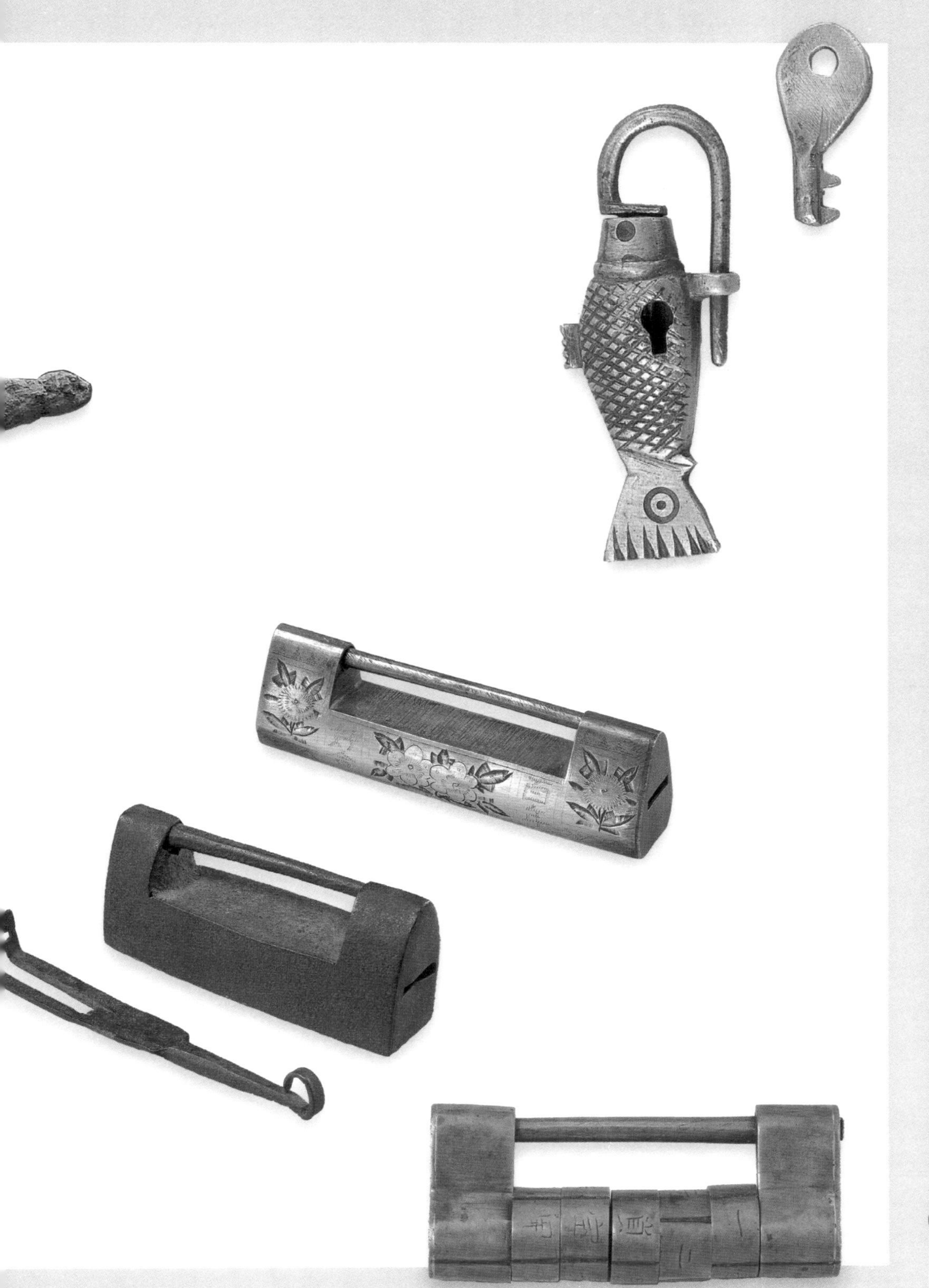

历史发展

HISTORICAL DEVELOPMENT

锁具的生成，与材料、工具及文化有密切的关联。观察锁具在历代的发展与使用，可看出当时的工艺技术水平、社会文化及经济发展程度。

最初，过着穴居生活的原始人类，为了防范野兽的侵袭与保护物品，会利用重石来挡住洞口，这可说是最早、最直接的原始安全装置。

随着材料的演进及用具的发展，安全装置的种类与功能逐渐增加与提升。有了绳索之后，先民为了保护贵重财物，常以精巧牢靠的绳结系紧，并设计出名为“觿”的兽牙来解开绳结。广义言之，“绳结”可说是中国最早的锁具，而“觿”则可说是中国最早的钥匙。当时亦有些锁具的设计，并无开关的功能，而是将其外观做成凶恶的动物形状，如老虎，借以防阻小偷，是一种象征性的吓人锁具。

The history of locks is in close association with the materials, tools, and cultural background of a specific time. And, the development and applications of locks in the past reflected the technological, cultural, and economical situations of each period in the history.

Primitive human beings that lived in caves learned to block the cave openings with heavy rocks to protect themselves and their belongings from the attack of beasts. This could very well be the earliest and the most direct primitive device for security.

With the advancement and development of new materials and tools, the types of security devices gradually increased and the functions upgraded. The Chinese ancestors tied knots with ropes solidly to secure their belongings and used a "*xi*" (tooth of the beast) to undo the knots. In broader terms, the knots were used as the earliest locks in ancient China and "*xi*" could be referred to as the earliest key. Some designs were not equipped with the substantial functions of locks for closing and opening. They were with patterns of the shapes of fierce-looking animals; for instance, a tiger, to scare the thieves away. It is a

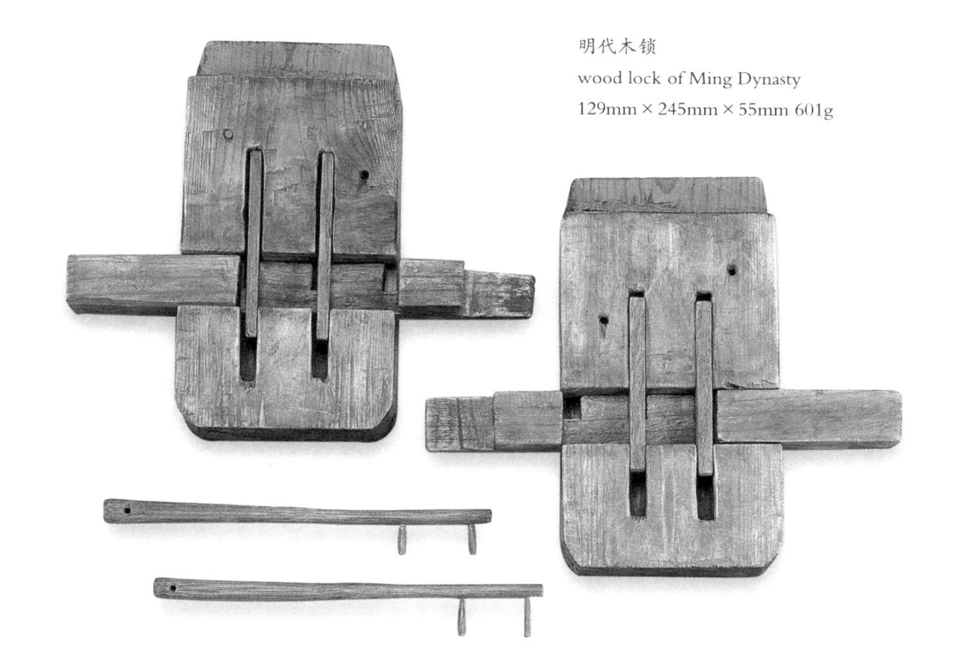

明代木锁
wood lock of Ming Dynasty
129mm × 245mm × 55mm 601g

木锁应是中国早期最具体的锁具，可追溯至石器时代的奴隶社会。据说迄今最早之木锁出土于仰韶文化（公元前 4000 ～公元前 3000 年）的遗址，但是现今并无真品存留，亦无正式文献加以记载。原始的木锁构造简单，门上有一个孔让竹竿类的横管式工具从门外进入，用以作用于门内的木栓来关门或开门。这亦是春秋时期 (公元前 770 ～公元前 476 年) 称钥匙为"管"或"籥"的原因。战国时代 (公元前 453 ～公元前 221 年) 的木锁有所改进，门上的木栓有一个圆孔，圆孔内装有上、下两根圆木棍，而钥匙则开始采用铜质的。

较为简单古拙的铜锁出现于青铜

"scaring lock" with only symbolic function.

Wooden locks should be the most substantial type of early Chinese locks. Its history can be traced back to the society of slavery in the Stone Age. It is said that the earliest wooden locks were found at the ruins of the Yangshao Culture (4,000 BC~3,000 BC). However, none of them is preserved until today, nor are they ever mentioned in official documents. The mechanism of primitive wooden locks is very simple. A hole on the door allows a pole-shaped tool to enter from outside to lock or unlock the wooden bolt on the inside. This also explains why the keys were called "*guan*" (tube) or "*yue*" (flute) in the Spring and Autumn Period (770 BC~476 BC). Wooden locks were much improved in the Warring States Period (453 BC~221 BC). A hole can be found on the wooden bolt that locks the door and two circular pegs were fixed horizontally in the hole. The keys at this time began to use bronze as the material.

汉代铜锁
brass lock of Han Dynasty
179mm × 37mm × 19mm 156g

器时代，锁内装有片状弹簧，利用钥匙与弹簧片的几何关系与弹力作用来上锁与开锁。到了春秋时代，锁具的设计开始复杂化，有些还装有机关。相传战国时代鲁班的改进，使得锁具得到普遍使用。

金属锁的大量使用，始于东汉末年（约公元 200 年），材料以青铜为主，并且出现了镂有虎、豹、麒麟、龟、蝴蝶等动物与昆虫造型的设计。唐代（公元 618 ~ 907 年）的制锁工艺相当发达，簧片锁的用途日益普遍，除了占多数的青铜制品外，有些则为黄铜、铁、银或金制品，其种类、外形及雕花亦日趋繁多。

再者，此时期的锁与钥匙，不乏美丽者，除富含艺术色彩之外，形态也相当多样化。此时期的锁与钥匙也是财富与权威的象征，达官显要与富贵人家，会在锁体刻上美丽的图案，有些更富含象征意义，但大部分的平民百姓，

Simple and plain bronze locks appeared in the Bronze Age. A thin, flat piece of spring can be found in the lock. The geometric relativity and the bouncing effect between the key and the spring were the mechanism for locking and unlocking. The design of locks got more complicated in the Spring and Autumn Period; some locks were designed with secret mechanism inside. And, some said that Lu Ban greatly improved Chinese locks for popular uses and mass production in the Warring States Period around 500 BC.

The application of large quantity of metal locks started in the late Eastern Han Dynasty around 200 AD, with bronze as the major material. And, some locks were designed with the patterns of animals and insects, such as tigers, panthers, qilins, turtles, and butterflies, etc. Locksmiths' technology level was pretty developed in Tang Dynasty (618~907). The usage of metal splitting spring locks, mostly made of bronze and partially of brass, iron, silver, or gold, was getting popular. And, the types, shapes, and engravings of locks were getting diversified.

In this period, some locks and keys were not only very beautiful and artistic colorful, but also with various shapes. Also, locks were one of the

宋代铁锁
iron lock of Song Dynasty
243mm × 30mm × 15mm 128g

钥匙 key
145mm × 12mm × 15mm 84g

仍只使用木锁。此外，当时的钥匙也被用来区分已婚与未婚的女人：所谓“出门带钥匙者”，是指已婚女子；而“出门未带钥匙者”，则是指未婚女子。

从汉代开始，金属簧片锁一直是中国人的主要用锁。两千多年来，中国传统锁具的外观虽然有所变化，但是内部构造始终没有太大的改进。到了20世纪40年代以后，由于西方栓销制栓锁的广泛使用，中国传统锁具才逐渐被淘汰。

symbols of wealth and power. Pretty pictures were engraved on the locks of the nobles and riches; some of them even bear totems. While in the same era, poor people used only wooden locks. Furthermore, in Tang Dynasty, a woman carrying a key in the outdoors indicated that she was married, and a woman carrying no keys meant that she was not married.

Ever since the Han Dynasty, metal splitting spring padlocks had always been the most widely used locks by Chinese people. Though the shapes of ancient Chinese locks diversified, the inner structures have not changed much for the past two thousand years. And, Chinese locks faded gradually after the western pin-tumbler cylinder locks were introduced into the country in the 1940s.

古锁语源

LINGUISTIC ORIGIN OF LOCKS

就现代的观点而言，锁是一种以钥匙、转盘、按键、电路，或者其他用具来操作的安全装置，用以防止物品被打开或移走，兼具防护、管理，甚至装饰的功能。然而，锁在中国历代的各种文献中，有着多种不同的称谓与定义。

《辞海》对“锁”的解释为“必须用钥匙方能开脱的封缄器”，《辞源》解释为“所以扃门户箱箧之具，使人不得开者，古谓之键，今谓之锁”，而《说文解字》则解释为“锁，铁锁，门键也”。

战国《周礼·地官·司门》中有“掌授管键以启闭国门”之句，老子《道德经》第二十七卷中有“善闭无关键而不可开”之句，西汉戴圣的《礼记·月令》中有“(孟冬之月)修键闭，慎管籥”之句。

In today's perspective, locks are security devices operated by keys, rotating plates, push buttons, circuits, or other means that are used to keep certain objects from being opened or taken away. And, locks are endowed with functions of protection, management, and even of decoration. Over the centuries, locks have been entitled to different names and definitions with literature from different dynasties in ancient China.

The dictionary *Cihai*[①] defines the lock as "a sealing device that requires a key to be opened". On the other hand, the dictionary *Ciyuan* defines the lock as "a device used to bolt doors and chests so to prevent people from opening them. It was named *jian* in the old times, now it is called *suo*." The book *Shuowen Jiezi* further explains: "*Suo*, iron locks, locks to the doors."

The *Si Men of Di Guan in Zhou Li* from the Warring States Period states: "He who is in charge of *guan* (tube) and *jian* (lock) opens and closes the gate of the state." The chapter 27 of *Dao De Jing* by Laozi reads: "Good shutting makes no use

① 为方便读者理解，本书部分英译书名用拼音表示。

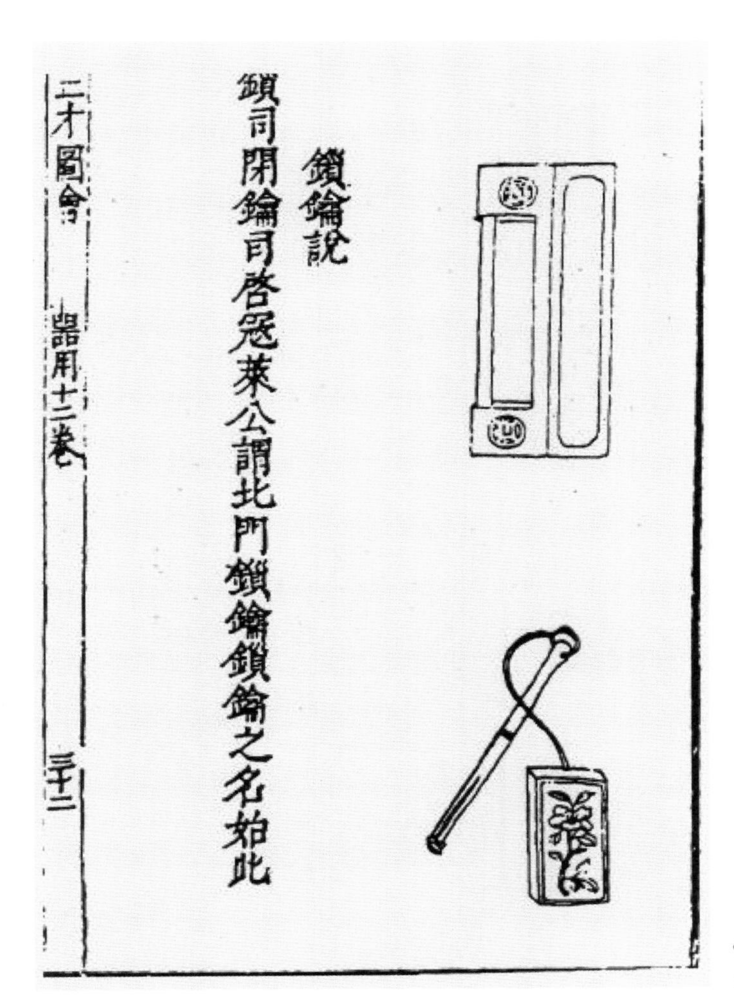

《三才图会》中的挂锁
padlock in *Sancai Tuhui*

《尚书·金縢》中有“启籥见书”之说，《辞源》解释道：“籥今通作钥。”汉代《芝田录》中有“门钥必以鱼”之说，明代张自烈的《正字通》解释道：“钥以闭户，匙以启钥。”《辞源》在解释“籥”字时说：“锁名，亦名钥。”

东汉班固于《汉书·小尔雅》中说：“键谓之钥。”而东汉许慎的《说文

《鲁班经》中的挂锁
padlock in *Lu Ban Jing*

of *jian* (lock), yet nobody can undo it." The *Yue Ling in Li Ji* by Dai Sheng from the Western Han Dynasty contains the phrase: "In the cold month of December, maintain the locks and be cautious with the keys."

The *Jin Teng of Shang Shu* states: "Open the *yue* (lock) and one sees the book." *Ciyuan* further interprets: "*Yue* is the same as what we called today *yao* (lock)." The book *Zhi Tian Lu* from the Han Dynasty contains: "Door locks are of fish-shaped." Zhang Zilie of the Ming Dynasty in his work *Zheng Zi Tong* further explains: "*Yao* is used to lock the door, and *shi* (key) to undo the *yao*." And, when *Ciyuan* explained the word *yue*, it says: "*Suo* (lock) also is named *yao*."

The *Chapter Xiao Er Ya of Han Shu* by Ban Gu in the Eastern Han Dynasty states: "*Jian* is also called *yao*." While Xu Shen of the Eastern Han Dynasty in his *Shuowen Jiezi* says: "*Qian* is a *suo* (lock)." Furthermore, Chen Pengnian and co-writers of the Song Dynasty in their work *Guang Shao* mentioned: "Soldiers seal doors with *qian* and have mighty axes ready for emergency."

Zheng Xuan of the Eastern Han Dynasty in his *Annotation of Yue Ling in Li Ji* wrote: "*Jian* (lock), the male part; and *bi* (keyhole), the female." The same idea was elaborated by Kong Yingda of the Tang Dynasty in his *Annotation of Yue Ling in Li Ji* as: "*Mu* (male) is the inserting part of locks, while *pin* (female) is the receiving part." *Ciyuan* made it clear that: "The keyhole on the door lock is called *bi*." Cai Yong of the Eastern Han Dynasty in his *Yue Ling Zhang*

解字》则解释道："钤，锁也。"宋代陈彭年等人的《广韶》中则有"兵钤以闭房，神府以备非常"之说。

东汉郑玄的《礼记·月令注》中说："键，牡；闭，牝也。"唐代孔颖达的《礼记·月令疏》解释道："凡锁器入者谓之牡，受者谓之牝。"《辞源》解释道："门闩之孔曰闭。"东汉蔡邕的《月令章句》中说："键，关牡也。所以止扉也，或谓之剡移。"意思是说牡是

Ju stated: "*Jian* is the sealing of *mu*. It is used to close the door, and also called *yan yi*. " This means that *mu* functions as the lock, therefore it can also be referred to as the door bolt.

Furthermore, Chinese characters " 籥 (*yue*)", " 钥 (*yao*)", " 管 (*guan*)", " 管籥 (*guan yue*)" were originated from fifes or flutes, and they also inherited the meaning of keys and locks.

Ancient locks with special functions and particular names were mentioned in some historical literature. For instance, the *Book of the Later Han* mentioned *Lang-Dang* lock(chain-and-shackle lock) as a padlock with metal

《农书》中的挂锁
padlock in *Book of Agriculture*

用作锁门的，因此也可以称为门闩。

此外，汉语中的“籥”“钥”“管”“管籥”代表着鼓笛或横笛，亦代表了锁与钥匙。

部分文献还记载有不同称谓与特别功能的古锁，如《后汉书》中提及“锒铛锁”，为结合金属链的挂锁，用以锁人犯。古锁也有因其外形而有特别称呼者，如10世纪时杜光庭的《录异记》中曾提及挂锁可称为“萎蕤锁”，因为此锁形同此种植物的管状根茎，且该锁含有连为一体的金属片，可自由压缩或伸张，故有此称谓。

根据古籍中所使用的称谓与所绘制的图案，以及出土与现存的古锁来判断，古早中国锁具应为机械式的挂锁。

chains for locking the prisoners. Some ancient locks were named after their peculiar shapes. In the tenth century, Du Guangting in his *Lu Yi Ji* mentioned a padlock called Polygonatum odoratum lock. The lock was so named probably because its appearance resembled the tubby root of polygonatum officinale, and the lock includes a compressible and extendable metal sheet.

Based on various names mentioned above, drawings in some literatures, and existing hardware, it is believed that an-cient Chinese locks should be mechanical padlocks.

古锁类型

TYPES OF LOCKS

古中国的挂锁大概可分为簧片构造锁与文字组合锁两大类。簧片（构造）锁使用时需要钥匙，又可分为广锁及花旗锁两种。（文字）组合锁使用时不需钥匙，只要将转轮上的文字转至正确的位置，即可开启。

Ancient Chinese padlocks can be classified into two categories: the splitting spring locks and the letter-combination locks. A splitting spring padlock has to use a key for opening, and it has the types of broad locks and pattern locks. A letter-combination padlock has no keys for opening, and it is unlocked when the letters of all wheels are rotated into the right positions.

Broad Locks

Broad locks are kinds of horizontal positioned locks, mostly used for the locking of doors, closets, chests, etc. *The Annotation of the Thirteen Classics* mentions that "East-west as wheel, and south-north as broad". The dictionary *Ciyuan* further elaborates that "North-south as vertical, and east-west as horizontal."

Most broad locks are made of bronze. And, the front side is of the shape of the character " 凹 ". The side end of broad locks before Ming Dynasty (1368~1644) appeared in the shape of a long circular tube. For broad locks in Ming Dynasty and Qing Dynasty (1644~1911), most were made of bronze and the shape of the upper side end is a triangle while the lower part is a slant rectangle. For broad locks after late Qing Dynasty, the shape of the upper side end is also a triangle and the lower part is a square. The majority of broad locks found in this era were made of brass, some made of Yunnan tutenag or iron. Most iron broad locks inherited the styles of the padlocks before the Ming Dynasty. Their side ends are of long circular-tube shape.

广锁

"广锁"是横式锁具，常用于锁门、锁柜、锁箱 等。"十三经注疏"中说"东西为转，南北为广"（一说"东西为广，南北为轮"），《辞源》解释为"南北为纵，东西为横"。

广锁的正面呈"凹"字状，大多为铜质。早期的广锁，其端面呈长圆筒形。中期的广锁，即明朝（1368～1644年）、清朝（1644～1911年）时期的广锁，其端面上部呈三角形、下部呈斜方形，材质大多为铜。晚清之后的广锁，其端面上部呈三角形、下部呈正方形，材质多为黄铜，亦有用云南白铜与铁制成者。大部分之铁质广锁沿袭了明代以前的式样，其端面呈长圆筒形。

花旗锁

“花旗锁”常用于锁抽屉、锁柜、锁箱等。“花”是指花样，“旗”则有表示的意思。《左传》中说“佩，衷之旗也”，“十三经注疏”中说“旗，表也”。

花旗锁具有不少的外形，可概分为人物、动物、乐器、字形、用品及其他类型，除了含有特定的寓意与吉祥之意外，亦兼具装饰功能。再者，花旗锁大多为铜质，镌雕洗练，工写兼蓄，做工精致传神，颇富民族传统色彩。

Pattern Locks

Pattern locks (*hua-qi* locks) were widely used for locking drawers, closets, and chests, etc. Exploring the linguistic origin of the pattern locks, “*hua*” means “floral pattern” and “*qi*” means "represent". *Zuo Zhuan* states: "Wearing it is representing it." *The Annotation of the Thirteen Classics* mentions: "*Qi* is the representation."

Pattern locks come in many different shapes. They can be roughly classified into the types of human figures, animals, musical instruments, letters, utensils, and others. The outer appearance of pattern locks is not just for special intentions and good fortunes, but also carries the purpose of decoration. Most pattern locks are made of bronze. The engravings on the locks are delicate, polished, and artistically fine though simple in style. The images engraved are vivid and alive. And, pattern locks are enriched with traditional civil arts and folk arts.

组合锁

组合锁通常具有三至七个转轮。锁体呈横式圆柱体形状，在圆柱之轴蕊上排列着数只同样大小且均能转动的转轮，每个转轮之表面蚀刻着同样数目的文字，只要所有转轮上的文字转到定位，且文字形成特定的词，就可开锁。

Combination Locks

Combination locks usually have three to seven wheels. They are of the horizontal round-pillar shape with several tunable wheels of the same size set in array on the central axis of the pillar body. Each wheel has the same amount of carved letters. Once the letters from all wheels are turned to the preset order or form a specific word, the lock is ready for opening.

古锁外形

SHAPES OF LOCKS

一般广锁的外形，大多为“凹”字形长方体，较无特殊变化。有些广锁是以锁体的外形来命名的，如早期的广锁呈长筒形，酷似爆仗，俗称“爆仗锁”；亦有广锁的外形呈方块状，故被称为“方锁”。此外，部分广锁的外形宛如一尾蜷曲的虾，因而名之为“虾尾锁”。

虾尾锁 shrimp-tail lock
115mm × 12mm × 10mm 28g

The majority of ancient Chinese broad locks appeared in the shape of Chinese character "凹" with little variation in this category. Some broad locks were named after the outer shapes of the locks. Broad locks in the early times adopted the shape of a firecracker; they were therefore called "firecracker locks". Some are square in shape and were called "square locks". Some broad locks took after the shape of a curling shrimp, and they were called "shrimp-tail locks".

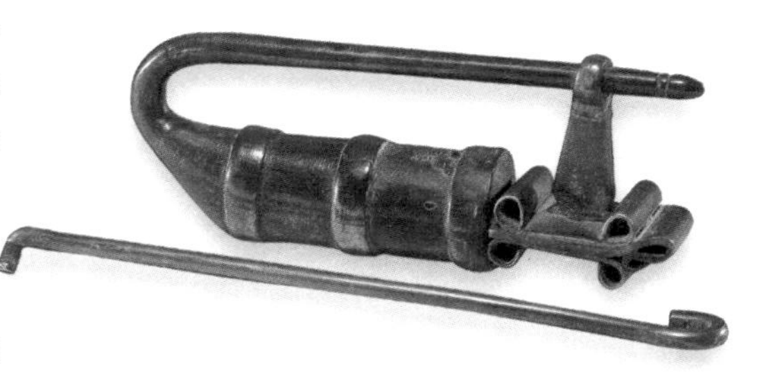

虾尾锁 shrimp-tail lock
100mm × 36mm × 20mm 100g

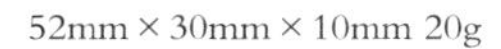

52mm × 30mm × 10mm 20g

64mm × 32mm × 12mm 41g

花旗锁 – 鱼 fish pattern locks

花旗锁的称呼，来自于其特殊且多样化的外形，除了具有保护、防护的作用之外，亦深具艺术价值。花旗锁最早、最多的外形为鱼形，汉代《芝田录》载有："门钥必以鱼者，取其不瞑目，守夜之义。"依此推断，鱼形的簧片挂锁，最迟出现于汉代（公元前 202 ~公元 220 年）。其后，随着时代的演进，花旗锁的外形逐渐多样化，有弥勒佛、八仙等人物造型，有琵琶、三弦琴、胡琴等乐器造型，有福、禄、寿、喜等字体造型，亦有葫芦、枕头造型。综观之，花旗锁的外形多为吉祥物，如鱼、龙、麒麟、蝙蝠、蝴蝶、虎、豹、马、狗、猴、乌龟、蝎等，不仅争奇斗艳，多姿多彩，同时上面大多还刻有特殊含义的吉祥图案。

65mm × 30mm × 15mm 44g

120mm × 37mm × 10mm 78g

77mm × 33mm × 10mm 59g

95mm × 36mm × 15mm 68g

Pattern locks get their names from the various, peculiar shapes of the locks. Not only they are equipped with the functions of protection and security but they are also of artistic value. Most of the pattern locks, especially the earlier ones, were made in the shape of fish. The book *Zhi Tian Lu* from the Han Dynasty mentions: "Door locks must take after the shape of fish, for fish sleeps with eyes open and so it guards the house at night." Therefore, the splitting spring padlocks with the shape of fish appeared no later than the Han Dynasty (202 BC~220 AD). With the progress of the society, the outer shapes of pattern locks became versatile. For example, the shape of Maitreya (a Buddhist messiah) and the shape of the Eight Immortals; the shapes of musical instruments, like the *pi-pa*, the three-string guitar, the *huqin*; the shapes of letters, "*fu* (good fortune), *lu* (richness), *shou* (longevity), *xi* (happiness)"; the shape of a bottle gourd or a pillow, etc. Many patterns locks came in fanciful shapes of lucky objects, such as fish, dragon, qilin, bat, butterfly, tiger, panther, horse, dog, monkey, turtle, scorpion, etc., all of them made the world of ancient locks more colorful and beautiful.

111mm × 54mm × 20mm 130g

182mm × 81mm × 23mm 564g

花旗锁 - 弥勒佛
maitreya pattern lock
83mm × 47mm × 26mm 198g

花旗锁－三弦琴
three-string guitar pattern lock
102mm × 23mm × 10mm 29g

花旗锁 – 胡琴
huqin pattern lock
98mm × 33mm × 40mm 115g

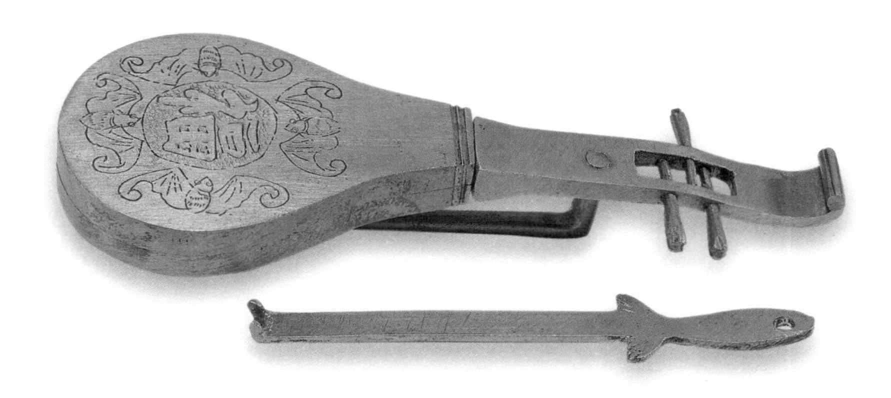

花旗锁 – 琵琶
pi-pa pattern lock
131mm × 48mm × 25mm 126g

花旗锁－福禄寿喜
letter pattern locks: *fu lu shou xi*
31mm × 22mm × 34mm 37g

花旗锁－葫芦
bottle gourd pattern lock
58mm × 104mm × 36mm 192g

花旗锁 - 龙
dragon pattern lock
98mm × 48mm × 16mm 104g

花旗锁 - 麒麟
qilin pattern lock
57mm × 48mm × 13mm 108g

花旗锁－蝙蝠
bat pattern lock
76mm × 49mm × 23mm 64g

花旗锁－鸟
bird pattern lock
95mm × 47mm × 16mm 80g

花旗锁－乌龟
turtle pattern lock
81mm × 44mm × 33mm 139g

花旗锁－蝎
scorpion pattern lock
145mm × 84mm × 62mm 265g

花旗锁－虎
tiger pattern lock
35mm × 48mm × 31mm 48g

花旗锁－猴
monkey pattern lock
28mm × 56mm × 14mm 50g

花旗锁 – 兔
rabbit pattern lock
43mm × 54mm × 12mm 44g

花旗锁 – 牛
ox pattern lock
83mm × 53mm × 18mm 159g

花旗锁 – 马

horse pattern lock

129mm × 85mm × 25mm 276g

花旗锁 – 狗

dog pattern lock

49mm × 33 mm × 12mm 38g

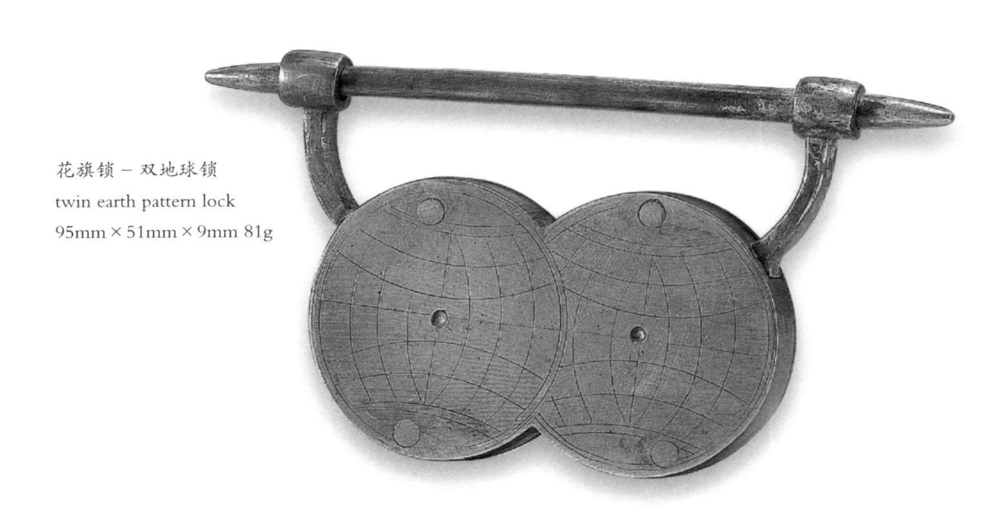

花旗锁－双地球锁
twin earth pattern lock
95mm × 51mm × 9mm 81g

花旗锁－炸弹
bomb pattern lock
48mm × 60mm × 15mm 47g

花旗锁 - 纪念钱币
commemorative coin pattern lock
44mm × 51mm × 12mm 44g

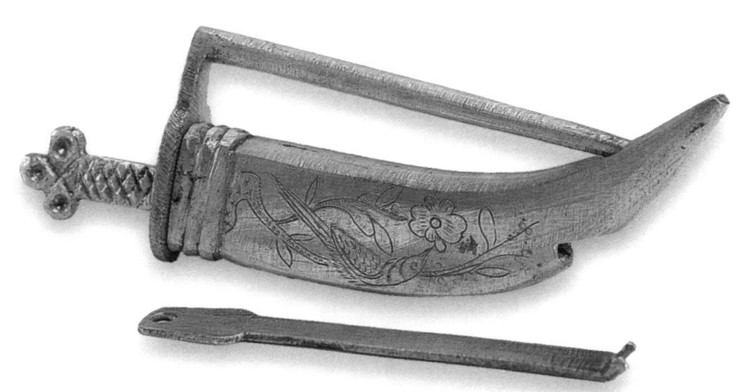

花旗锁 - 弯刀
machete pattern lock
127mm × 46mm × 13mm 108g

百家锁 *bai-jia* lock
120mm × 29mm × 29mm 31g

百家锁 *bai-jia* lock
112mm × 28mm × 28mm 48g

百家锁 *bai-jia* lock
37mm × 11mm × 11mm 8g

有种古锁的外形呈圆柱状，体似水桶，锁梁上部像水桶把手，底部拖个尾巴，锁体表面刻有“百家保锁”四个字，属民俗器物，材质为金、银、铜等。相传此种锁具乃是民间百姓生儿育女时，向百户邻家募款请工匠打造而成，因此俗称“百家锁”。百家锁有多种样式，大多悬挂在家中或小儿的颈项上，用以消灾祛邪，保佑孩子长命百岁。

"*Bai-jia* lock" or "hundred-family lock" is a civil art handicraft with several shapes, made with gold, silver, or bronze.Chinese characters "*bai* (百) *jia* (家) *bao* (保) *suo* (锁)" were engraved on the lock body, that means lock of protection from a hundred families. It is said the family with a newborn baby raised the money from one hundred families in the neighborhood to hire a locksmith to make the lock. A hundred-family lock is usually of pillar shape with the lock body the shape of a barrel and the shackle of the sliding bolt the shape of barrel handle with a tail at the bottom of the lock. It is hung in the house or on the neck of the baby to keep away the evil spirits and disaster and pray for the baby's longevity.

锁体雕花
ENGRAVING OF LOCKS

锁体的雕花有镂刻与蚀刻两种，常见的图样有吉祥物、人物、文字、山水、花草及其他事物，不仅拙中藏巧、朴中显美，亦显示出古中国特有的风格与民族语言。主要的吉祥物有龙、凤凰、麒麟、鹤、鹊、蝴蝶及蝙蝠，另有鱼、狮、虎、百合、灵芝、萱草、荷花、芙蓉、梅等。

Engravings on the body surface of ancient Chinese locks can be classified into two types: the etching and the engraving. Patterns commonly employed are lucky objects, human figures, Chinese characters, landscapes, flowers, plants, and others. All these revealed hidden handicraft skills and great beauty in an object of such utility. In this sense, locks are an excellent example of unique ancient Chinese habits and social language.

There are many lucky objects in ancient China, mainly the dragon, phoenix, qilin, crane, magpie, butterfly, bat; fish, lion, tiger, so are lily, ganoderma, day lily, lotus, hibiscus, plum blossom, etc.

吉祥纹饰
auspicious patterns

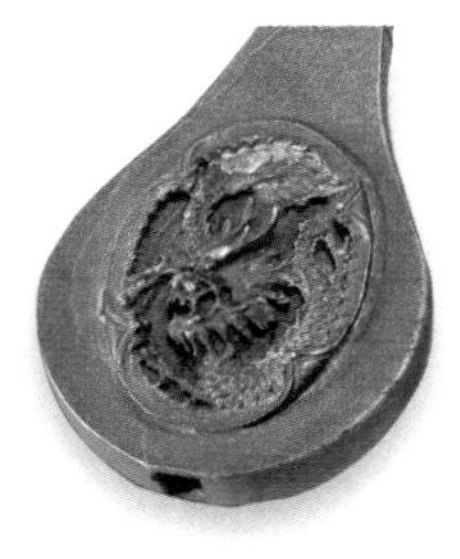

镂刻－凤与龙
etching: phoenix and dragon

蚀刻 – 双龙戏珠

engraving: two frolic dragons with a ball

195mm × 55mm × 24mm 569g

蚀刻 – 麒麟

engraving: *qilin*

149mm × 38mm × 24mm 275g

蚀刻 – 鱼

engraving: fish

66mm × 50mm × 17mm 134g

蚀刻 – 狮

engraving: lion

112mm × 40mm × 23mm 251g

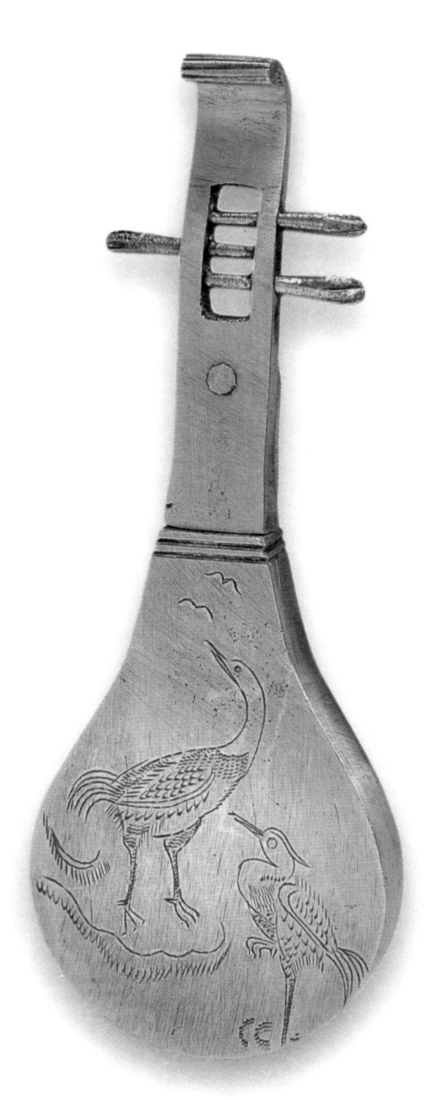

蚀刻 – 鹤
engraving: crane
47mm × 130mm × 25mm 114g

蚀刻 — 鹊

engraving: magpie

47mm × 130mm × 25mm 120g

福 禄 寿

锁体上的人物有的为福禄寿三星，有的为美丽女子，也有的镂刻着和合二仙，栩栩如生，饶富神韵。

Some of the patterns engraved on the lock body are the three lucky gentlemen—"*Fu-Lu-Shou*", beautiful women, and the two harmonic fairies—"*He-He*". All were lively and vividly engrave.

蚀刻 – 福禄寿三仙

engraving: three lucky gentlemen—*Fu-Lu-Shou*

150mm × 67mm × 30mm 558g

蚀刻 - 童子嬉戏
engraving: children at play
140mm × 47mm × 36mm 409g

蚀刻 - 童子嬉戏
engraving: children at play
179mm × 48mm × 34mm 614g

蚀刻 - 人物
engraving: human figures
80mm × 32mm × 21mm 108g

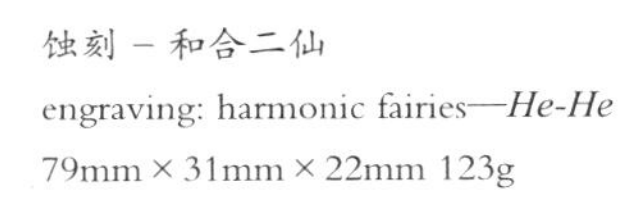

蚀刻 - 和合二仙
engraving: harmonic fairies—*He-He*
79mm × 31mm × 22mm 123g

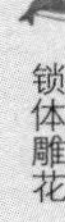

蚀刻－词句
character engraving: expression
118mm × 40mm × 23mm 163g

刻于锁体上的文字鲜明地反映出当时人们的期望，如福如东海、功名百代、状元及第、五子登科、五子三元、五世其昌、百子千孙、金玉满堂、梅开五福、红梅结子、万事如意、一本万利、百代千秋、同心永爱、百年好合、如月之恒、福、禄、寿、喜等。部分则刻有诗词，蕴涵浓郁的文艺气息。此外，有时亦可据此得知古锁的制造年代与人名。再者，锁体上的花草、祥云、山水、房舍等图案，十分生活化。

The letters and words engraved on the locks also revealed the expectations of the people in the old times, for instances, "happiness as deep as the east China sea", "success and fame for a hundred generations", "ranked top in the national examinations", "all five sons succeed in the government examinations", "all five sons ranked top three in the government examinations", "prosper for five generations", "blessed with many descendents", "full house of gold and jade", "plum blooms with lots of happiness", "red plum flowers bear seeds", "all the wishes come true", "making countless profits", "lasting for a hundred generations and a thousand years", "joined hearts with eternal love", "sound and well for a hundred years", "eternal as the moon", "good fortune", "richness", "longevity", "happiness", etc. And, some locks were artistically engraved with poetry that instilled in the locks a literary spirit. Furthermore, sometimes the owners, manufacturers, and the time of the manufacturing year could be traced back through the wordings. Other patterns observed are true-to-life topics like flowers and plants, fair cloud, natural scenery, houses, etc.

文字蚀刻 – 五子登科

character engraving: all five sons succeed in the government examinations

156mm × 44mm × 21mm 310g

文字蚀刻 – 百子千孙

character engraving: blessed with many descendents

165mm × 49mm × 30mm 442g

文字蚀刻 – 金玉满堂

character engraving: full house of gold and jade

117mm × 29mm × 24mm 260g

文字蚀刻 - 五子三元

character engraving: all five sons ranked top three in the government examinations

39mm × 40mm × 14mm 61g

文字蚀刻 - 梅开五福

character engraving: plum blooms with lots of happiness

100mm × 72mm × 20mm 200g

文字蚀刻－百年好合

character engraving: sound and well for a hundred years

125mm × 37mm × 22mm 211g

文字蚀刻－万事如意

character engraving: all the wishes come true

150mm × 69mm × 23mm 245g

文字蚀刻 – 红梅结子

character engraving: red plum flowers bear seeds

123mm × 42mm × 25mm 268g

文字蚀刻 – 如月之恒

character engraving: eternal as the moon descendents

57mm × 30mm × 17mm 66g

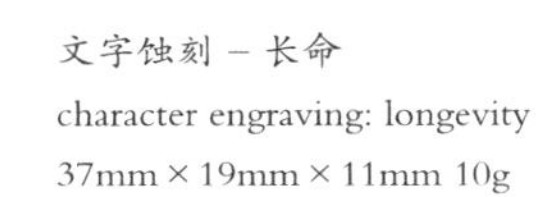

文字蚀刻 – 长命
character engraving: longevity
37mm × 19mm × 11mm 10g

文字蚀刻 – 永保千秋

character engraving: stay the glory a thousand years

54mm × 38mm × 27mm 30g

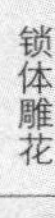

文字蚀刻 – 福

character engraving: good fortune

55mm × 55mm × 18mm 55g

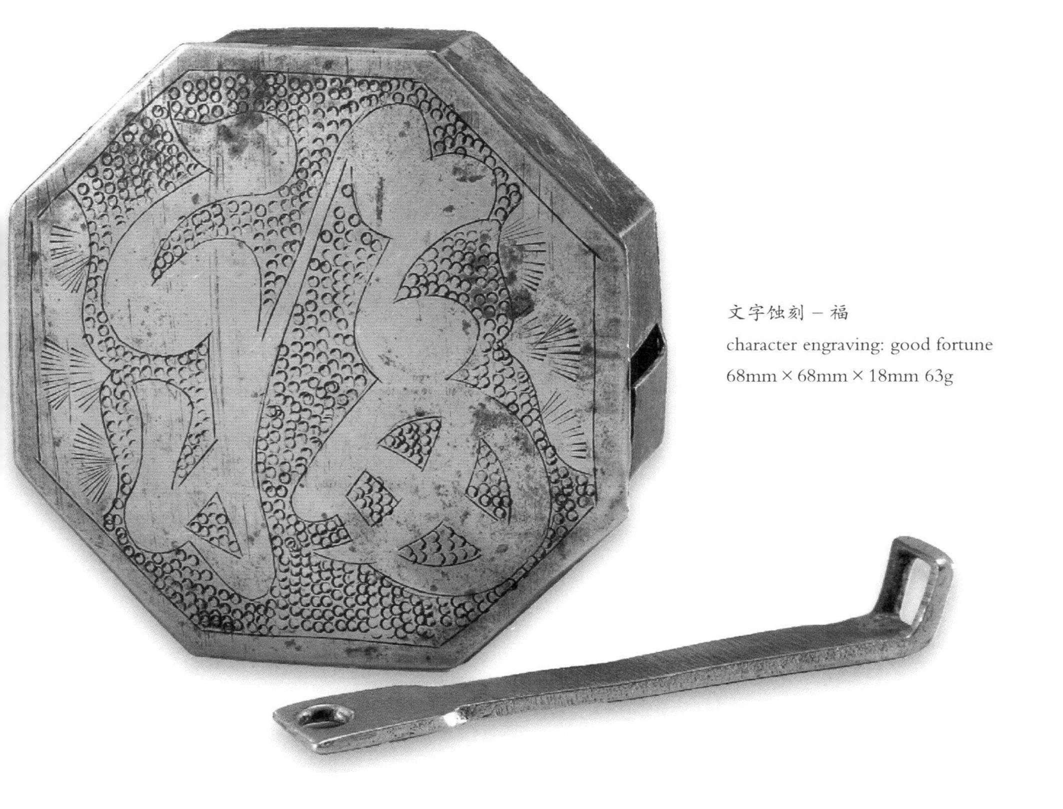

文字蚀刻 – 福
character engraving: good fortune
68mm × 68mm × 18mm 63g

文字蚀刻 – 福
character engraving: good fortune
69mm × 38mm × 16mm 103g

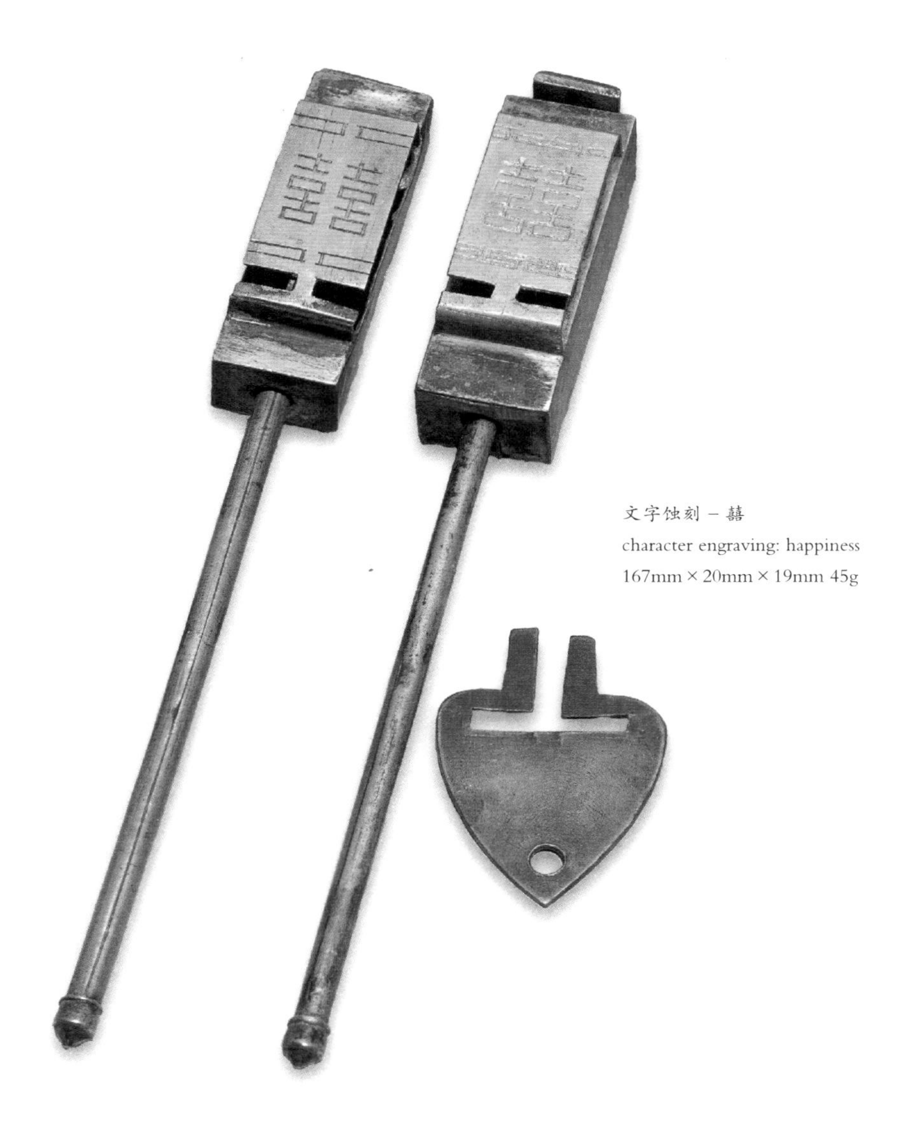

文字蚀刻－囍
character engraving: happiness
167mm × 20mm × 19mm 45g

文字蚀刻 – 囍

character engraving: happiness

88mm × 31mm × 22mm 126g

诗词蚀刻

character engraving: poetry

156mm × 46mm × 25mm 435g

花草蚀刻
engraving: flowers and plants
79mm × 32mm × 19mm 115g

花草蚀刻
engraving: flowers and plants
216mm × 45mm × 26mm 560g

花草蚀刻

engraving: flowers and plants

89mm × 33mm × 20mm 157g

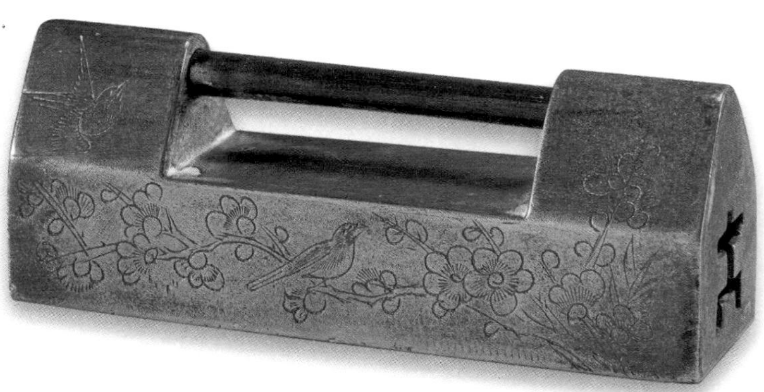

蚀刻 - 梅鹊

engraving: magpie and blossom

77mm × 29mm × 20mm 103g

花草蚀刻

engraving: flowers and plants

41mm × 85mm × 18mm 45g

蚀刻 – 纹饰

engraving: decorated totem

75mm × 43mm × 19mm 98g

蚀刻 – 制造工坊

engraving: manufacturer's shop

锁体材质

MATERIALS OF LOCKS

材料在古中国的发展，左右了制作锁具的材质。古中国在石器时代以石头、骨头、木头作为工具的材料。在公元前4000年，即生产出性能远优于自然铜的黄铜。在公元前8世纪左右，出现了第一批高炉，是人类最早大规模炼铁的开始。在公元前4世纪炼出了白铜。在5世纪时则发明了灌钢。

根据不同时代、不同材料的发展，古锁的材质有木材、青铜（铸造）、黄铜（铸造）、红铜（铸造）、白铜、铁（熟铁、锻铁、锻铁和烙铁）、银、金、钢、铝及镍等。早期的广锁以青铜材质最为流行，后期的广锁大多为黄铜材质，有些为铁质，而像银般的（云南）白铜，外观优美，质感华丽，广为豪门巨室、达官显要所喜爱，亦有少数讲究者使用景泰蓝材料。

The materials for manufacturing locks have been mainly affected by the historical development of materials in ancient China.

In Stone Ages, stones, bones, and woods were the materials for making tools. Brass appeared in around 4,000 BC that was superior in its quality than the blister copper. The large scale of ironmaking started with the first appearance of blast furnace in around 8th century BC. Cupro nickel was produced before 4th century BC. And, the process of steel casting was invented in around 5th century AD.

According to the development of various materials in various periods, ancient Chinese locks were made of wood, bronze (casting), brass (casting), red bronze (casting), cupro nickel, iron (wrought iron, forged iron, forged and soldered iron), silver, gold, steel, aluminum, and nickel. The early broad locks found were mostly made of bronze; later the brass was the most popular, followed by iron. Cupro nickel that looks like silver with its beautiful, quality look was the favorite of noble families, high-rank officials, and rich businessmen. Some went so far as to use enamel to serve as the material of the locks.

木锁 wood lock

129mm × 245mm × 55mm 601g

青铜锁
bronze lock
121mm × 41mm × 22mm 217g

黄铜锁
copper lock
16mm × 46mm × 16mm 11g

红铜锁
red brass lock
263mm × 72mm × 42mm 1758g

白铜锁
cupro nickel lock
123mm × 40mm × 28mm 318g

铁锁

iron lock

88mm × 34mm × 17mm 83g

银锁

silver lock

40mm × 49mm × 7mm 44g

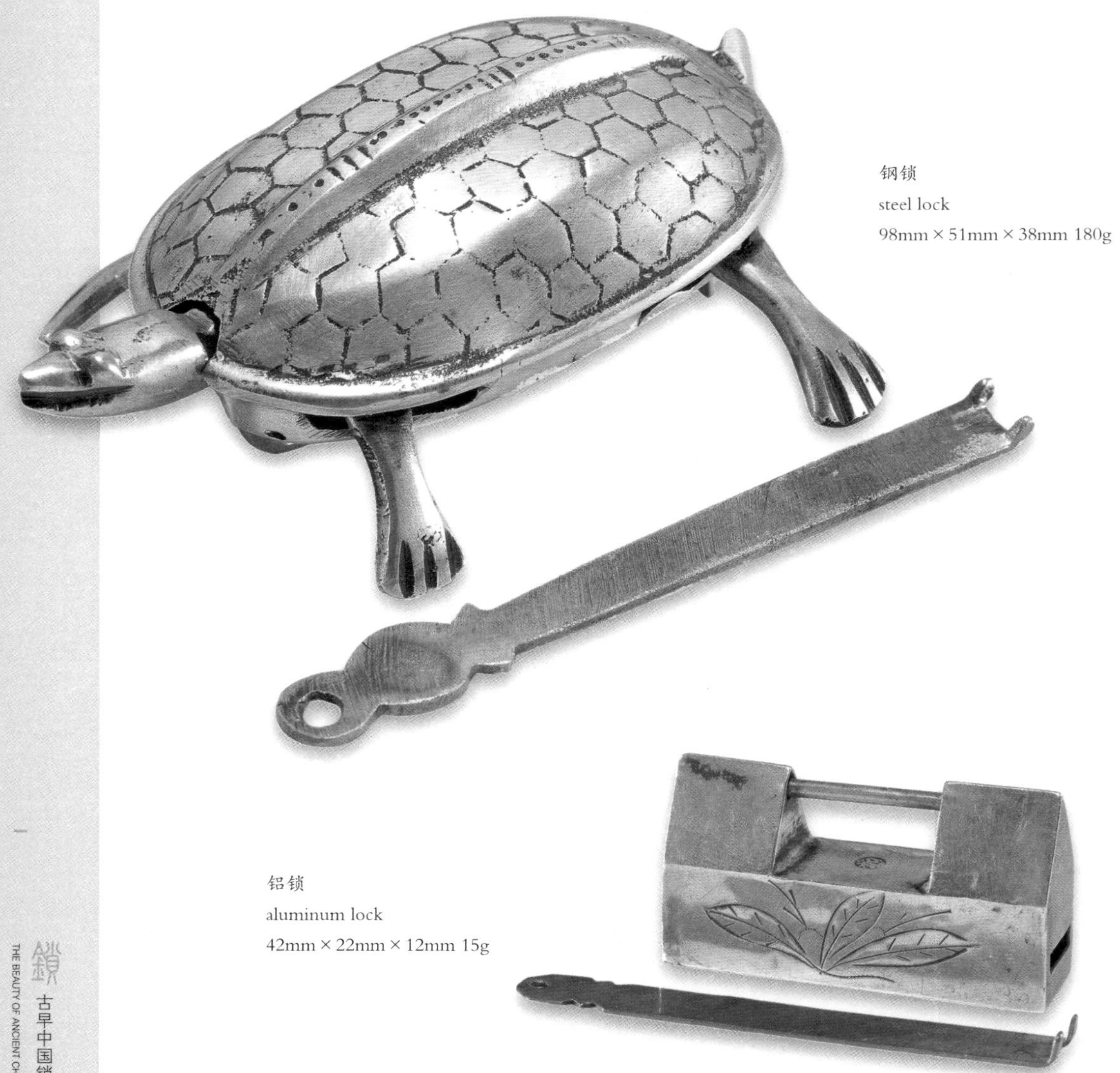

钢锁
steel lock
98mm × 51mm × 38mm 180g

铝锁
aluminum lock
42mm × 22mm × 12mm 15g

镍锁

nickel lock

150mm × 62mm × 15mm 219g

景泰蓝锁

enamel lock

46mm × 25mm × 13mm 30g

古锁构造

MECHANISMS OF LOCKS

簧片锁为古中国最典型的锁具，由锁体、锁栓（具有锁梁与分离弹簧片）和钥匙组成。锁体提供了钥匙孔，让钥匙插入，并导引锁栓滑动。锁栓的一部分为锁梁，用以挂锁，另一部分为栓梗，用以固结分离弹簧片的一端。钥匙则是根据钥匙孔的位置与形状及弹簧片的构形而设计。上锁时，锁栓上的弹簧片因弹力的作用而张开，弓卡在锁体的壁内。开锁时，钥匙头恰可挤压钳制张开的弹簧片，使锁栓滑动与锁体分离。由于簧片锁是利用簧片弓卡在锁体壁内而上锁的，所以亦称为“撑簧锁”。

The most typical ancient Chinese lock is the splitting spring lock. It consists of a lock body, a sliding bolt, and a key. The lock body provides a keyhole for the key to insert and the supporting guide for the sliding bolt to move. The sliding bolt has a shackle for hanging the lock and a stem for bonding one end of the splitting springs. The key is designed corresponding to the configuration of the splitting springs, and the location and shape of the keyhole. When it is locked, the sliding bolt is trapped by the opening splitting springs against the inner wall of the lock body. For opening, the key is inserted and its head squeezes the opening splitting springs so that the sliding bolt can be separated from the lock body.The splitting spring lock is also called "prop-open spring lock" because it functions with the trapping and propping up of springs against the inner wall of the keyhole.

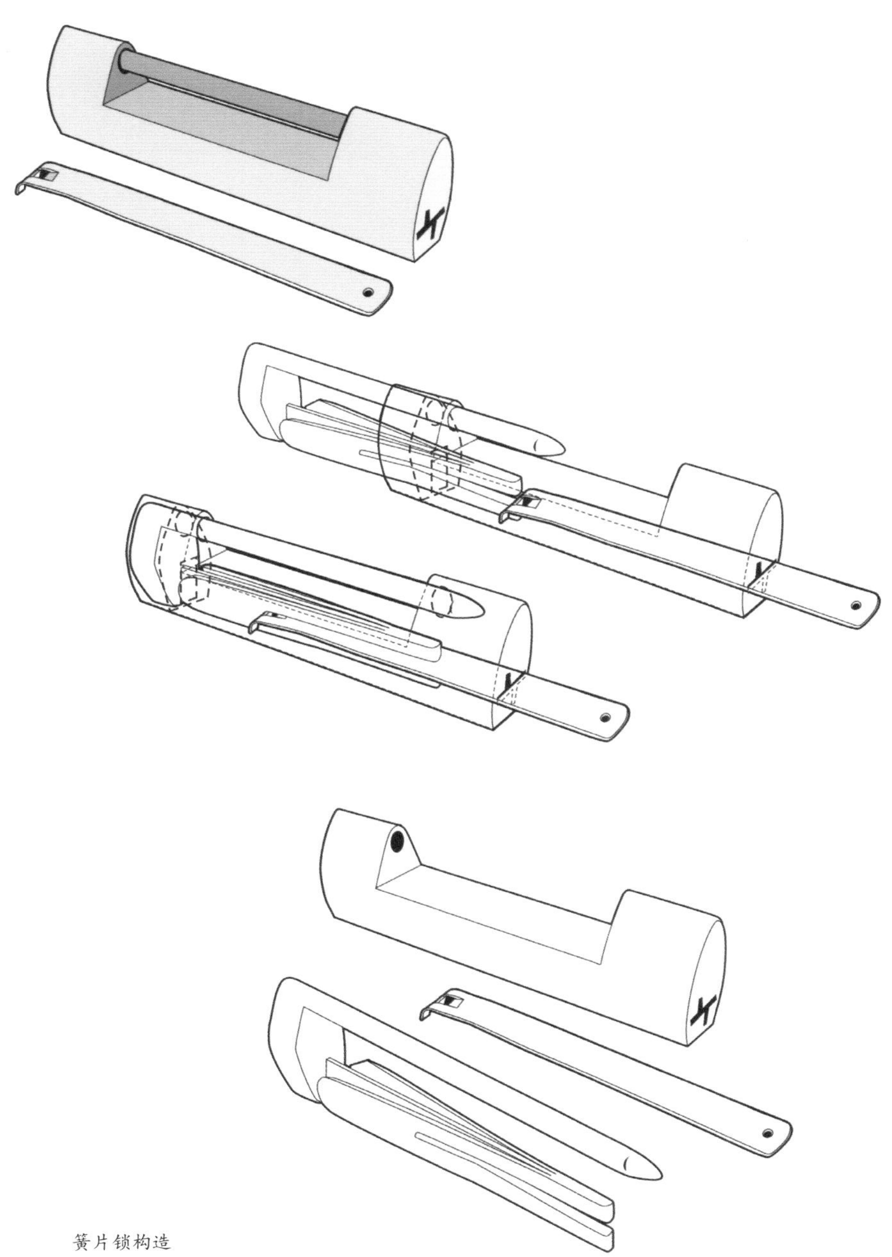

簧片锁构造

mechanism of splitting-spring lock

簧片锁

splitting-spring lock

97mm × 46mm × 15mm 199g

黄片锁

splitting-spring lock

文字组合锁
letter-combination lock
90×52×17mm 146g

组合锁由锁体、转轮和具有锁梁的锁栓组成。锁体包括一个片状端板与转轴，让转轮转动，并导引锁栓滑动。锁栓亦有一个片状端板，一部分固结锁梁，用以挂锁，另一部分固结具有凸片的栓梗。每个转轮的大小一样，表面大多镂刻着四个文字，其内径并有一凹形槽与栓梗上的凸片对应。开锁时，先将所有转轮上的文字，在锁体的正面排成一条线，且形成特定的字符串，使所有转轮的凹形槽向上对齐，构成一个通道，此时便可滑动锁栓、与锁体分离，锁便被打开了。

A combination lock comprises of the lock body, rotating wheels, and the sliding bolt with a shackle and a stem. The lock body contains an end plate and an axis with rotating wheels for guiding the movement of the sliding bolt. The sliding bolt also has an end plate for bonding both the shackle to hang the lock and the stem with several convex (凸)-shaped blocks. Every rotating wheel is of the same size. Usually four letters are engraved on the surface. And, there is a concave (凹)-shaped chute that corresponds with each convex-shaped block on the stem. When unlocking the lock, one has to rotate the letters on each wheel into the correct order and position. When all the concave-shaped chutes face upward, a channel is formed that allows the stem with convex-shaped blocks to slide apart from the lock body. The lock is then opened.

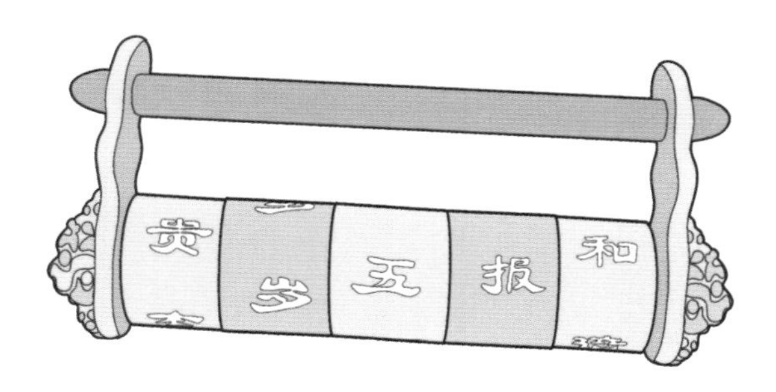

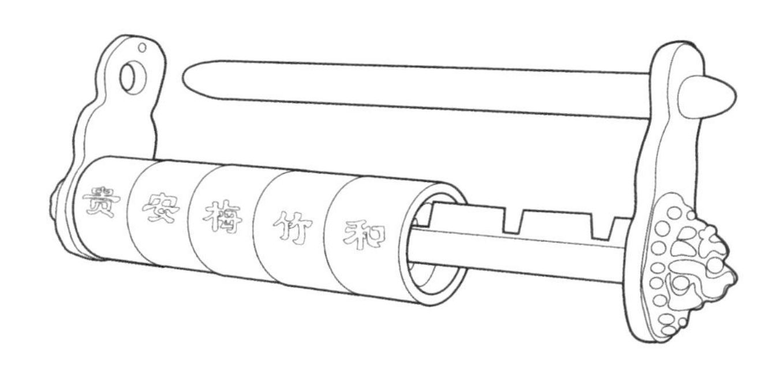

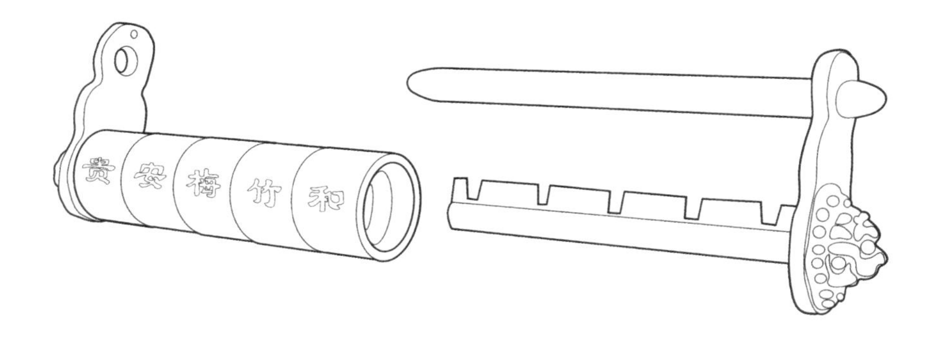

文字组合锁构造

mechanism of letter-combination lock

簧片构形
CONFIGURATIONS OF SPRINGS

簧片锁的主要特征为锁栓上的分离弹簧片构形，包括弹簧片的类型、数目、位置及大小，不仅影响了钥匙头与钥匙孔的设计，而且关系着锁具的开启方式。弹簧片为金属长方形薄片，有四种类型。弹簧片的数目大多为二至六片。弹簧片的一端固接在栓梗的尾部，另一端为张开状，呈对称排列。同一锁具的弹簧片，大多具有相同的大小，有些则有不同的长度。

One of the major characteristics of a splitting spring padlock is the configuration of the splitting springs on the sliding bolt. It includes the design of the type, the number, the arrangement, and the size of the springs. These factors affect not only the design of the key-head and the keyhole, but also the approach for opening the lock. A splitting spring is a thin piece of rectangular metal, and generally of four types. The numbers of the splitting springs are normally ranged from 2 to 6. One end of the splitting springs is fixed to the stem of the sliding bolt and the other end is open for trapping against the inner wall of the keyhole. These splitting springs are normally arrayed symmetrically, usually all of the same size. However, in some special designs, splitting springs are different in length.

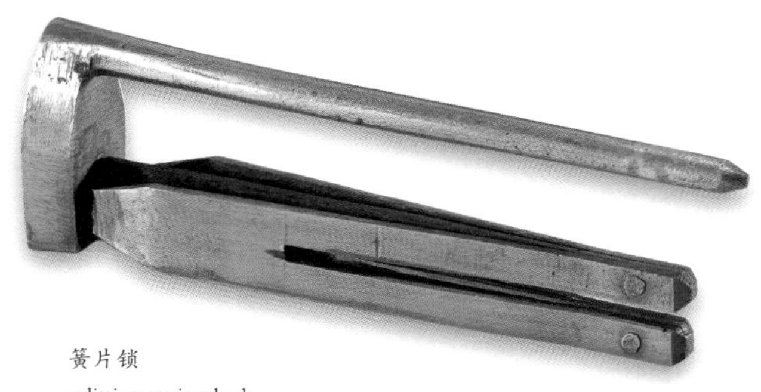

簧片锁
splitting-spring lock

二片簧

TWO-SPRING

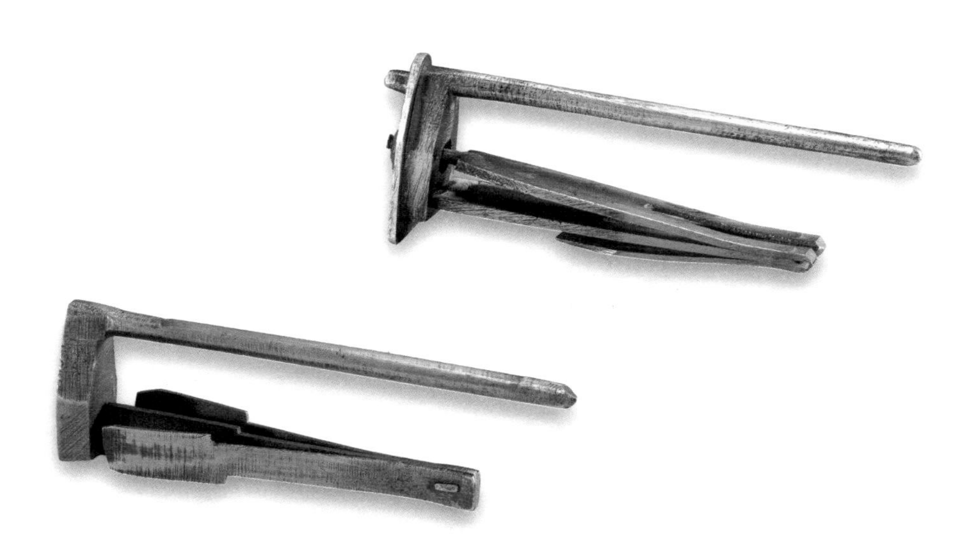

三片簧

THREE-SPRING

相同的弹簧片构形，可能有不同的挤压钳制方式，即有着不同设计的钥匙头，而相同的挤压钳制方式，并不意味着有相同设计的钥匙与钥匙孔。汉代的簧片锁大多只有三片弹簧，因此俗称“三簧锁”。

Although some locks have the same configuration of splitting springs, they may have different ways of squeezing, i.e., different designs of key-heads. And, configurations of splitting springs with the same squeezing may have different designs of keys and keyholes. Most of the splitting spring locks that appeared in Han Dynasty were designed with three splitting springs. So, they are commonly known as "three spring locks".

四片簧

FOUR-SPRING

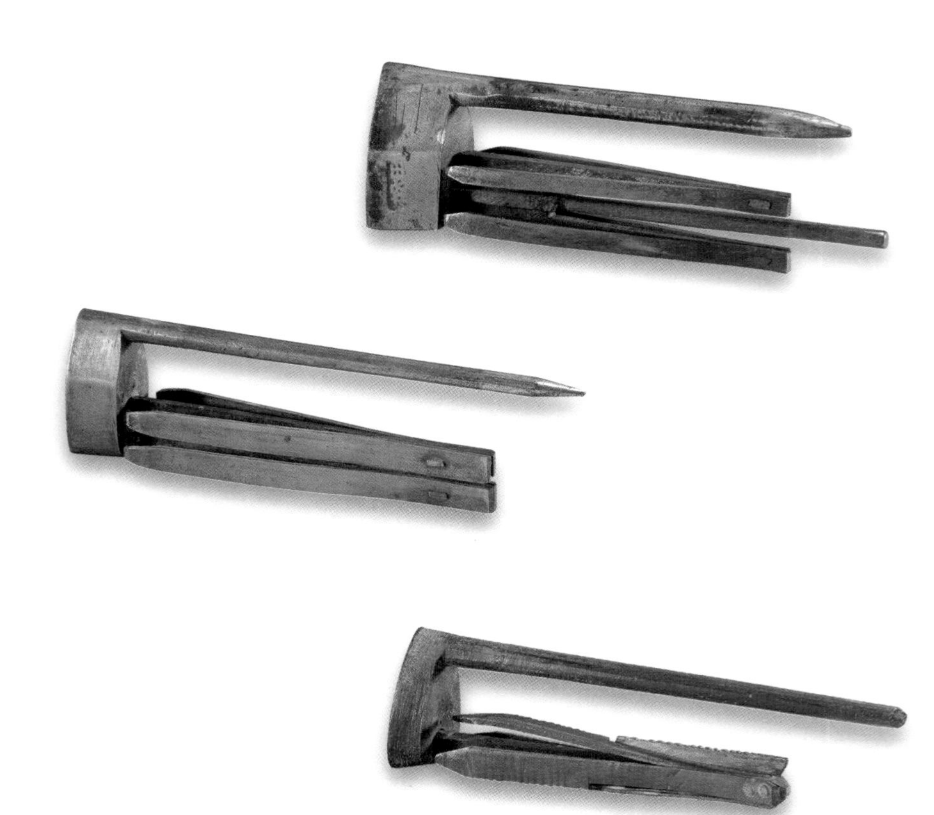

五片簧

FIVE-SPRING

钥匙

KEYS

中国古代锁具的钥匙可依其数目来分类：使用一把钥匙来开锁者，其钥匙称为简单钥匙；使用两把以上的钥匙来开锁者，其钥匙称为复合钥匙。大部分的古锁以简单钥匙即可开启，有些古锁则必须利用复合钥匙方可打开。

古锁的钥匙大多配有一个空白的钥匙胚，用以保护钥匙，并可用来打造备份钥匙。此外，钥匙胚亦有助于锁的开启。有些锁具的锁体较长，其钥匙虽然能够完成开锁的动作，但在施力或拔出时会感不便，钥匙胚的存在，可发挥伸长钥匙的作用，方便锁具的开启。

Keys of ancient Chinese locks can be classified into simple keys and compound keys, according to the number of keys needed for opening the locks. Most ancient Chinese locks were designed to open by one key, i.e., by a simple key. Some were special designed to have two or more different keys, i.e., a compound key, for opening the lock.

In most cases, a key blank is provided with either a simple key or a compound key to protect the key and/or for preparing a spare key. Moreover, a key blank is helpful for the opening of locks with longer lengths. It can double the length of a key, and thus helps pushing the sliding bolt for opening or pulling out of the key from the springs.

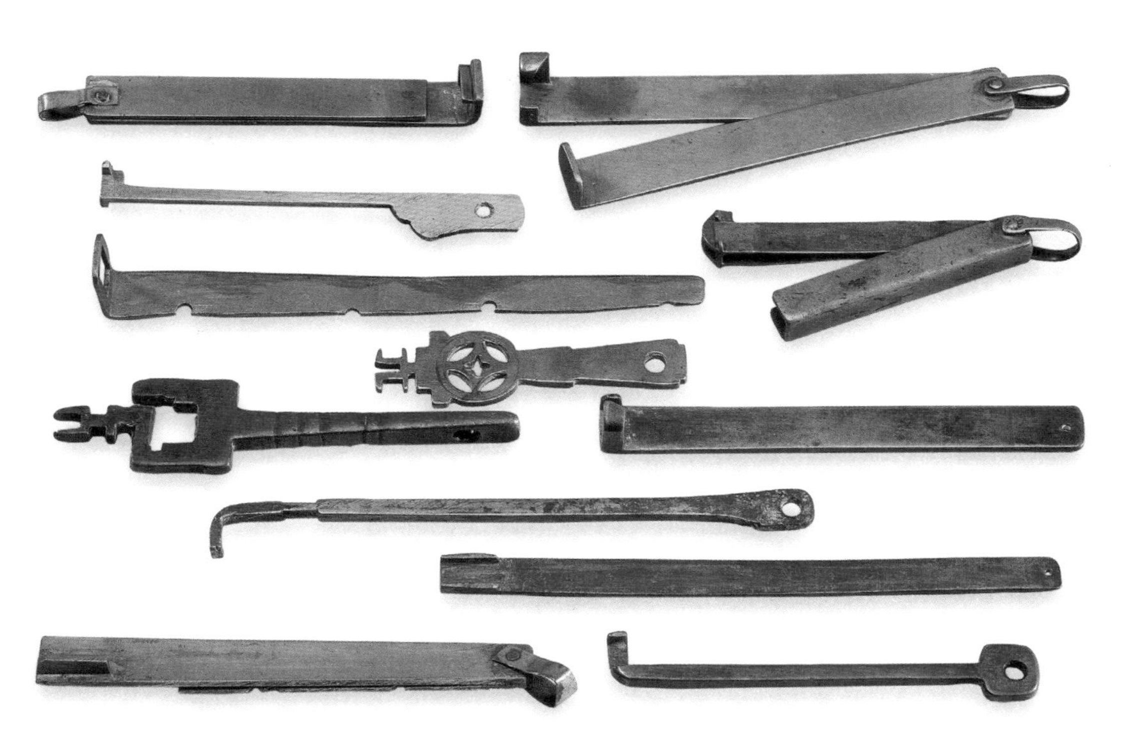

简单钥匙
simple key

复合钥匙
compound key

钥匙与钥匙胚
key and key blank

簧片锁的钥匙头的形状非常多样，有平面的，亦有立体的，其设计是为了让钥匙插入锁孔并配合弹簧片的构形挤压钳制簧片以开锁。平面钥匙头的形状，可分为完全包合、不完全包合及不包合三种。立体钥匙头的形状则不一，但挤压钳制弹簧片的方法大多采用不包合的设计。

The key-heads of splitting spring locks could be either of planar shapes or spatial shapes for the purposes of entering the keyholes and squeeze on the splitting springs to open the locks. There are numerous shapes of key-heads for various ancient Chinese locks. The shapes of planar key-heads include total-enclosure design, partial-enclosure design, and non-enclosure design. Spatial key-heads come in various shapes. However, non-enclosure designs are normally used for squeezing the splitting springs.

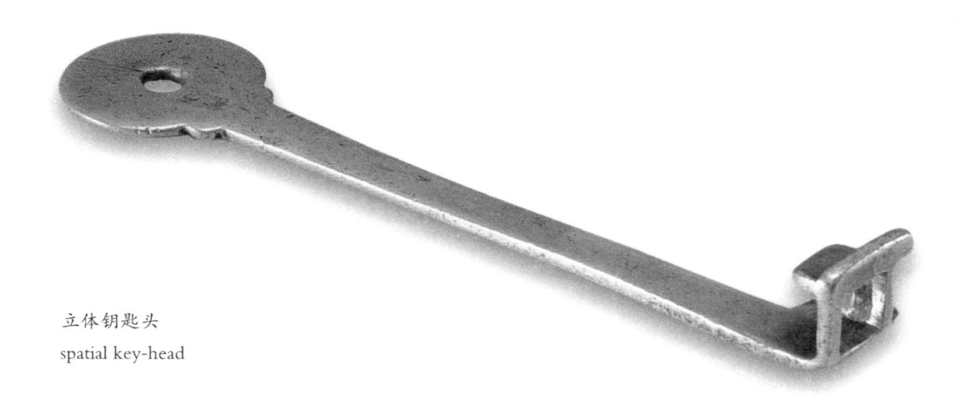

立体钥匙头
spatial key-head

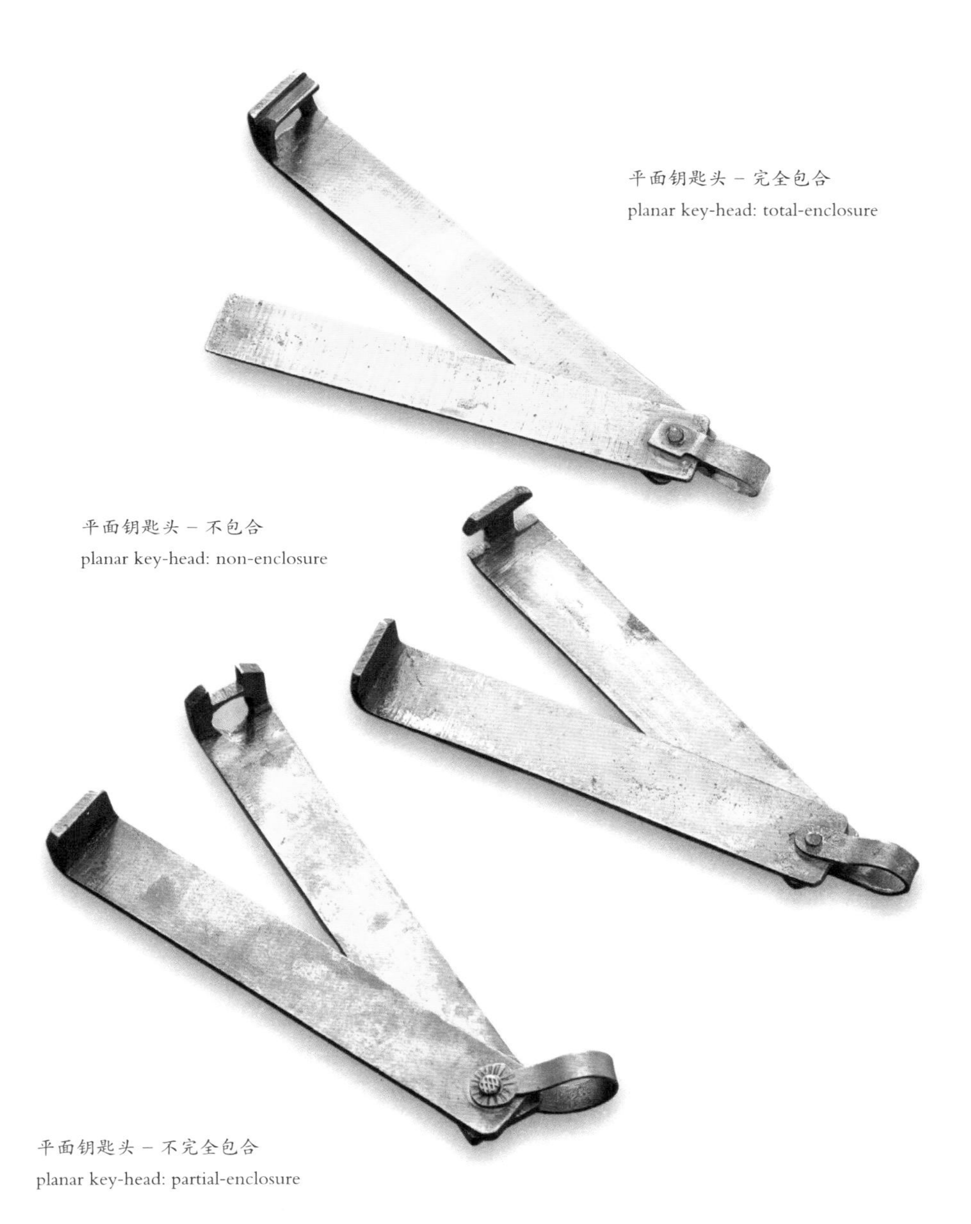

平面钥匙头 – 完全包合
planar key-head: total-enclosure

平面钥匙头 – 不包合
planar key-head: non-enclosure

平面钥匙头 – 不完全包合
planar key-head: partial-enclosure

钥匙孔

KEYHOLES

簧片锁是利用钥匙来开启的，因此必须有钥匙孔让钥匙插入。钥匙孔的构形（即位置与形状）设计，基本上是为了使相配对的钥匙插入以开锁，并防止不配对的钥匙进入。

An ancient Chinese key-operated padlock should have a keyhole for the key to insert for opening. The configuration, i.e., the position and shape, of a keyhole is basically designed for the corresponding key to enter and for preventing a foreign key from opening the lock.

位置

钥匙孔的位置，大部分开在锁体的端面，有些开在正面、背面、上面或底面，有些则同时开在相邻接的数个面上。若钥匙孔仅出现在锁体的某一面，则称为平面钥匙孔，否则，即称为立体钥匙孔。立体钥匙孔的位置与形状的巧妙设计，使古锁形式越发多样化，因难以将钥匙插入钥匙孔中，往往令人持钥兴叹，给予开锁人智慧的考验。

Positions

A keyhole is normally on the sidewall of the lock body. Sometimes it appears on the front side, the backside, the topside, or the bottom side. And, some keyholes are positioned across different adjacent sides of the lock body. If a keyhole is on one side end of the lock body, it is called a planar keyhole; otherwise, it is a spatial keyhole. The design of the positions and shapes of spatial keyholes enriches the types of ancient locks. The difficulties of inserting the key into the keyhole, which pose a challenge for the people's wisdom, always make people sigh.

钥匙孔位置

positions of keyholes

大部分的钥匙孔是开放式的，有些则是隐藏式的。开放式钥匙孔可直接由锁体表面找到。隐藏式钥匙孔大多以一底板、端板覆盖，必须先找到开启底板、端板的方法，再滑动底板、旋转端板以显现出钥匙孔。简单的设计，只要移动锁体底部的平板，即可顺利地找到钥匙孔。复杂的设计，则必须先按动锁体上的某一按钮，以些微拉开锁栓，才能转动端板，并滑动底板以显现出钥匙孔。

有些古锁是以其钥匙孔的位置来命名的，如钥匙孔开在背面者称为“背开锁”，钥匙孔开在上面者称为“上开锁”。

Most ancient Chinese splitting spring padlocks have open keyholes and keys can be inserted directly to open the locks. However, some locks were designed with hidden keyholes. A hidden keyhole is usually covered with a plate. The plate has to be found first. Then, the plate should be rotated or slid to reveal the keyhole. For a simple design, one can slide the bottom plate of the lock to reveal the keyhole. For a complicate design, a button on the lock body should be located and pressed to release the locking pin of the end plate first. Then, the end plate can be rotated for the bottom plate to slide to reveal the keyhole.

Some ancient locks were named according to the positions of the keyholes. For examples, the keyhole of a "back-opening lock" appears in the backside of the lock body, and the keyhole of a “top-opening lock” appears in the topside of the lock body.

开放式钥匙孔
open keyhole
62 × 38 × 27mm 119g

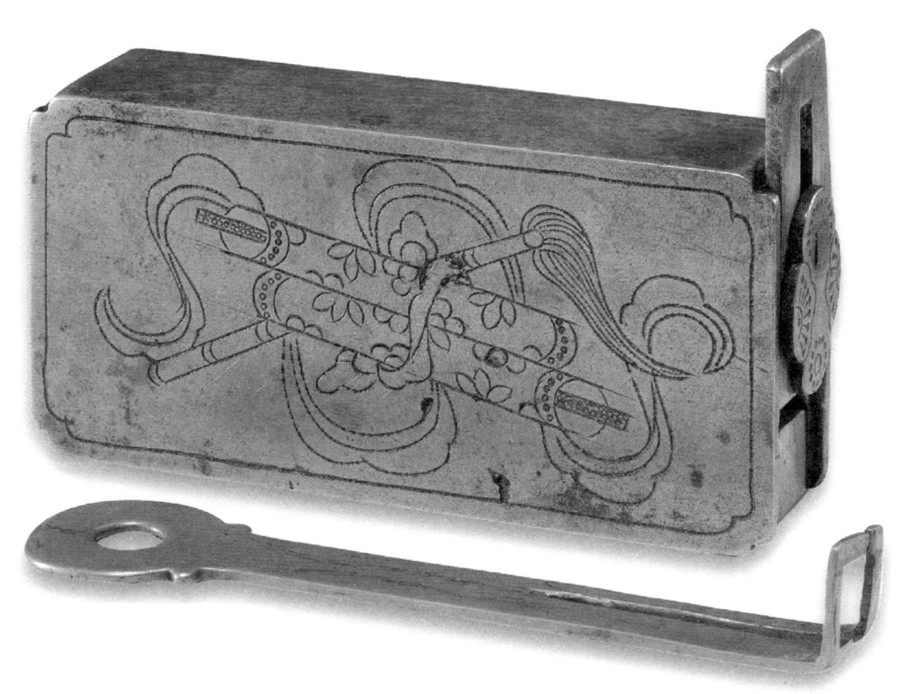

隐藏式钥匙孔

hidden keyhole

82mm × 42mm × 19mm 251g

形状

无论是平面钥匙孔或者是立体钥匙孔，其形状都相当多样，除了配合钥匙头的构形设计外，有些还因使用者的社会地位而异。

广锁的外形虽然大同小异，但钥匙孔的形状却有着多种形式，其复杂度按官职递增，代表了当时社会的阶级制度。再者，制锁工坊皆按官府的规定行事，不能僭越。

位于锁体端面钥匙孔的形状，大部分为文字形，如“一、上、下、土、工、士、山、古、而、吉、尚、喜、寿”等。孔状为“一”字者，是庶民百姓所使用的锁具；为“士”字者，是读书人、士大夫所使用的锁具；为“吉”字者，是达官贵人所使用的锁具；“喜”字用于婚嫁，“寿”字则用于祈求健康、长寿。而帝王、将相、太子、王公、皇妃、公主等，则另有标志。

有些古锁是以其钥匙孔的形状来命名，如钥匙孔呈“一”字形者称为“一字锁”，钥匙孔呈“士”字形者称为“士

Shapes

Both the planar and the spatial designs of keyholes come in many shapes. They were designed to match the configuration of the key-heads and the social status of the owners.

Although most of the broad locks are more or less identical with each other, the shapes of keyholes varied greatly with many patterns. The higher rank the owner's social status was, the more complex the design of the keyholes became. Furthermore, locksmiths designed and manufactured locks according to the rules of the administration court, no transgression was allowed.

Some ancient Chinese splitting spring padlocks were designed with letter-shape keyholes on the side end of the lock body, such as " 一、上、下、土、工、士、山、古、而、吉、尚、喜、寿 ", etc. They were made for users of various social classes. The shape of Chinese character " 一 " was used by the civilians, " 士 " was used by intellectual class, and " 吉 " was used by the nobles and high rank officials. The shape of Chinese character " 喜 " was used in weddings, and the shape of Chinese character " 寿 " was used to pray for the bliss of good health and long life. Furthermore, there are significant shapes of keyholes for various ruling classes such as the

钥匙孔形状

shapes of keyholes

定向锁钥匙孔

keyhole of fix-orientation lock

86mm × 40mm × 20mm 189g

字锁”，而钥匙孔呈“寿”字形的锁具，则称为“寿字锁”。寿字锁的钥匙为一个寿字或两个寿字，利用寿字的构形挤压钳制弹簧片的作用开启，设计独特，用来祝颂万寿无疆、福如东海、寿比南山。

有些具有立体钥匙孔的广锁被称为“迷宫锁”或“定向锁”，其钥匙孔位于底面接近端面的直角处，开凿成“工”字形。这类锁具的特征是：其钥匙不容易直接插入钥匙孔。

emperors and kings, generals and ministers, the oldest princes, royal families, queens, princes and princesses,etc.

Some ancient locks were named after the shapes of their keyholes. For instances, the locks with keyholes of the letter " 一 " shape was called the " 一 letter locks"; the locks with keyholes of the letter " 士 " shape were called " 士 letter locks". Therefore, " 寿 letter locks" are locks with keyholes of the letter " 寿 " shape. The key of a " 寿 letter lock" was usually designed as one " 寿 " letter or double—" 寿寿 " letter. This is a unique and complicated design with a key-head the shape of the letter " 寿 " to squeeze on the splitting springs. Such a lock was often presented on birthdays as a gift of good wishes on endless happiness and long life.

Some broad locks with spatial keyholes are called "labyrinth locks" or "fix-orientation locks". They have their keyholes located at the vertical corner formed by the bottom side and one side end of the lock body with the shape of a " 工 " letter. The main characteristic of this type of locks is that it is not easy to insert the key into the keyhole.

古锁开启

OPENING OF LOCKS

一般而言，簧片锁需钥匙才能开启，而组合锁则不需钥匙，只要将转轮上的文字旋至正确的位置即可开启。簧片锁的设计，是利用简单的原理构思出巧妙的开启方式，可分为以下四个步骤：找到钥匙孔的位置，将钥匙插入钥匙孔，将钥匙置于开锁的位置，将锁具打开。对于一些特殊的设计而言，如何找到钥匙孔的位置，是一种挑战。找到钥匙孔的位置之后，如何将钥匙插入钥匙孔，是一门学问。再者，就算得以进入钥匙孔，也要懂得如何转折，才能将锁具打开。

拥有隐藏式钥匙孔之古锁的装置十分巧妙，锁体的两个端面上各装有一个轮形花样摘子（旋钮），呈对称状，其中一面的摘子是固定的，另一面的摘子则是活动的，转动活动摘子才可移动锁梁、打开盖板，才得见钥匙孔。这

Generally speaking, splitting spring locks take keys to open while letter-combination locks can be open without a key.The principle of designing ancient Chinese splitting spring padlocks is simple. However, the opening mechanism is sometimes ingenious. It normally includes the following four steps: find out the keyhole, insert the key into the keyhole, move the key to the final position, and open the lock. For some special designs, it is a challenging work to find the keyhole, it is difficult to insert the key into the keyhole, and it is sometimes not even easy to move the key to the right final position for opening the locks.

Some locks with hidden keyholes are very delicately designed. Both side ends of the lock body have an attached wheel-shaped button. One is fixed and the other is movable. When the movable one is lid upon, the sliding bolt could then be moved to enable the covering plate to be opened and the keyholes then revealed. Such locks are named "four-open locks", since they take four actions to open the locks. Such locks are also named "puzzle locks", since the keyholes are hidden so well such that many puzzles have to be

类锁具，由于开锁须分四次进行，因此被称为“四开锁”。再者，由于必须打开多重机关才能找到钥匙孔，因此又被称为“机关锁”。有些古锁的钥匙孔隐藏在底面，其上盖有镂成花样的铜盖板，必须滑动铜盖板才会显露出钥匙孔，此类锁具被称为“花边锁”。有些古锁的钥匙孔为锁体与锁栓间的缝隙，钥匙为一薄片金属，用以从缝中插入开锁，此类锁具被称为“暗门锁”。

solved before finding the keyholes. Some locks have their keyholes hidden behind the bottom side and covered by etched copper plates. The plate has to be slid aside to reveal the keyhole. Such locks with hidden keyholes are named "pattern-side locks". Some locks with hidden keyholes are named "hidden door locks"; the keyhole of such a lock is actually a seam between the lock body and the sliding bolt that can be inserted and unlocked by a key with the shape of a thin piece of metal.

四开锁
four-open lock
93mm × 30mm × 20mm 146g

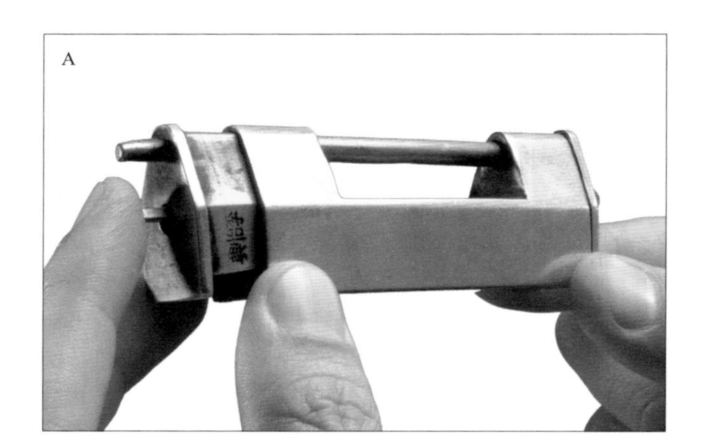

古锁开启步骤
steps of open lock
ABCDE

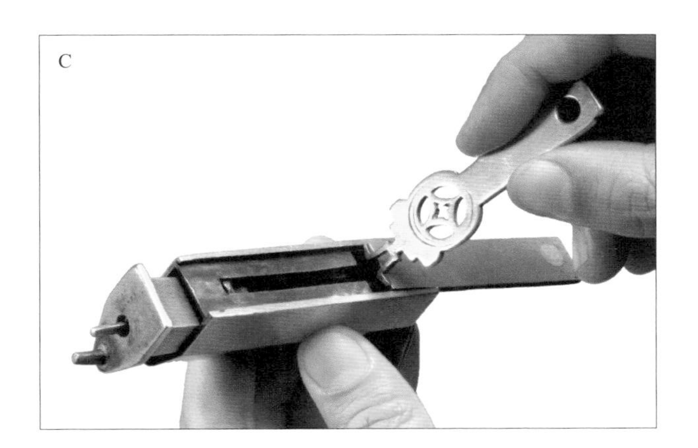
C

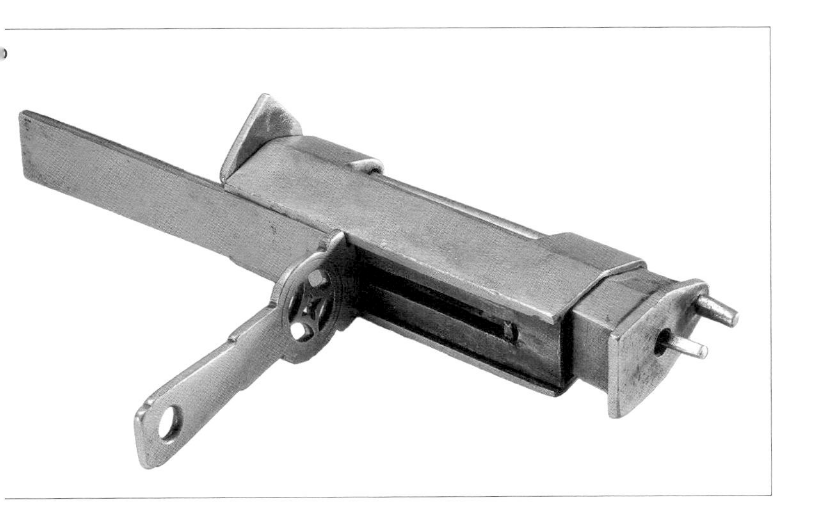
D

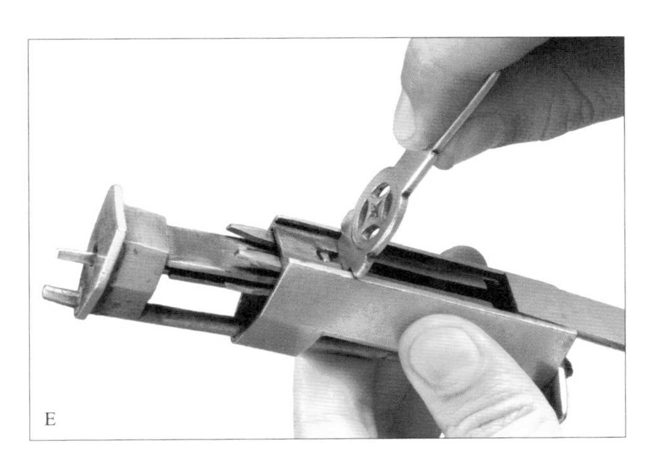
E

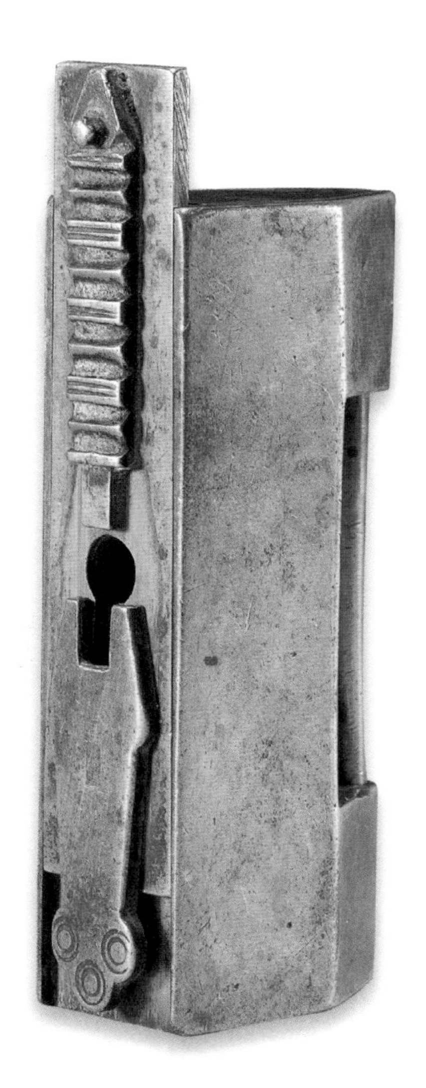

花边锁

pattern-side lock

78mm × 31mm × 20mm 166g

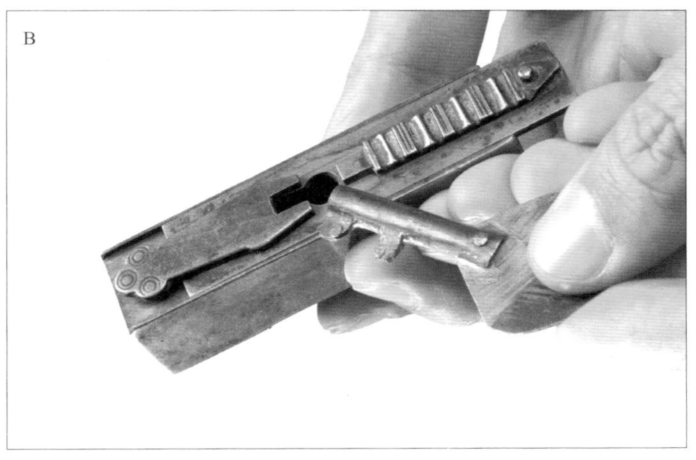

花边锁开启步骤

steps of opening a pattern-side lock

AB

将钥匙插入钥匙孔的设计，和钥匙头与钥匙孔的构形有关，可为简单式插入，亦可为复合式插入。简单式的设计，可将钥匙直接插入钥匙孔。复合式的设计，必须将钥匙头的特定部分在特定的方位与锁孔的特定位置接触，才能将钥匙插入，这类锁具被称为“迷宫锁”或“定向锁”。这样的设计，即使他人拥有正确的钥匙，一时之间也难以将锁具打开。

有些设计简单的古锁，钥匙插入钥匙孔之后即处于开锁位置，称为单段式开锁，而有些设计复杂的古锁，须将钥匙以适当的方向分别推动、转动，才能到达开锁的位置，此称为多段式开锁。

The design of the keyhole depends on the configuration of the keyhead and the keyhole. The insertion of keys can be either simple insertion or complex insertion. For a simple design, the key can be inserted into the keyhole directly. For a complex design, the right portion of the key-head must contact the keyhole in the right position and in the right orientation in order to enter the keyhole, and such locks are named "labyrinth locks" or "fix-orientation locks". In such a design, even if a stranger has the right key, it is difficult for him to insert the key into the keyhole for engaging the splitting springs.

For simple designs, namely one-stage open, once a key is inserted into the keyhole it reaches the final position for opening the lock. For complicate designs, namely multiple-stage open, however, a key has to be pushed and/or rotated in several right directions, respectively, to reach its final position.

定向锁
fix-orientation lock
79mm × 31mm × 18mm 108g

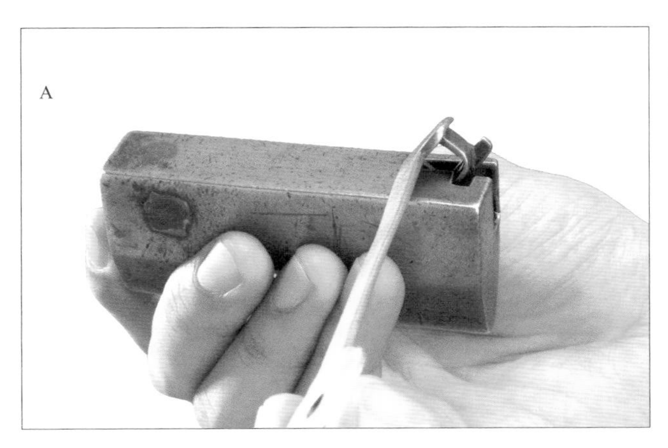

定向锁开启步骤

steps of opening a fix-orientation lock

ABC

当钥匙插入锁体处于开锁位置之后，接下来的动作为直推或旋转钥匙，此时钥匙头会将张开的弹簧片挤压钳制，使锁栓与锁体分离以开锁。再者，钥匙插入钥匙孔后要旋转方能开启的锁具，被称为“转冲锁”。

有些古锁的栓梗上装有长、短两种弹簧片，需要两把不同的钥匙来开锁。上锁时，长簧片卡在锁体壁内，开锁时，必须先用一把钥匙挤压钳制长簧片，使锁栓移动，直到短簧片卡在锁体壁内，接着用另一把钥匙挤压钳制短簧片，使锁栓整个通过锁体壁内的卡口而开锁，这类锁具被称为“二开锁”或“双开锁”。

Once the key is inserted into the lock in the right position, the last step for opening the lock is to push or rotate the key for the key-head to squeeze the splitting springs to separate the sliding bolt from the lock body. Some locks are named "rotating-open locks" if they are finally opened by rotating the keys.

Some multiple-stage-open locks are sometimes called "double-open locks". They have long and short splitting springs attached to the stem of the sliding bolt, and they take two different keys to be unlocked. When it is locked, the lock is trapped by the longer springs. When people try to open the lock, one key has to be inserted first to squeeze the long springs to shift the sliding bolt until the lock is trapped by the shorter springs. Another key is then used to squeeze the short springs to enable the sliding bolt to be entirely released.

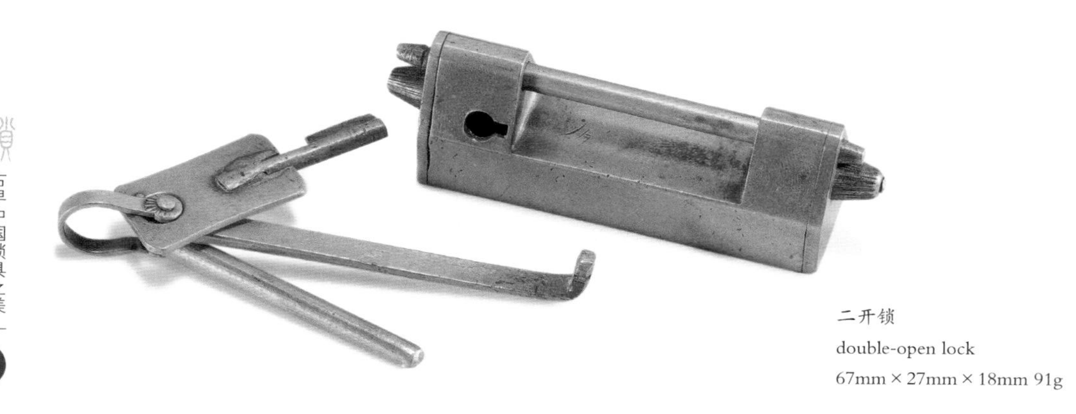

二开锁

double-open lock

67mm × 27mm × 18mm 91g

转冲锁

rotating-open lock

88mm × 31mm × 22mm 126g

有些设计特殊的簧片锁，是利用摩擦力将钥匙倒拉至插销的扣上来开启，这类锁具被称为“倒拉锁”。再者，亦有些设计特殊的簧片锁，不用分离的钥匙来开启，如利用插销卸扣的特殊动作来开启，这类锁具被称为“无钥锁”。

Some locks are named "pullback locks". Once the key is inserted into such a lock, it is unlocked by pulling the key back against the buckle of the sliding bolt through frictional force. Furthermore, some special designed splitting spring locks can even be unlocked without separated keys, and they are named "keyless locks". For example, one type of keyless locks is opened with the special movement of the shackle of the sliding bolt.

倒拉锁

pullback lock

87mm × 31mm × 19mm 107g

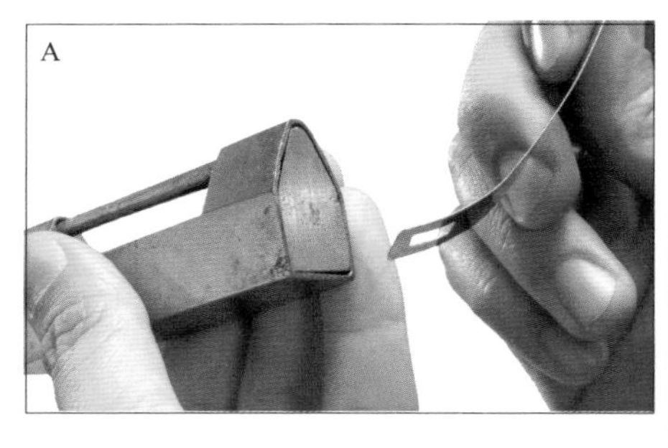

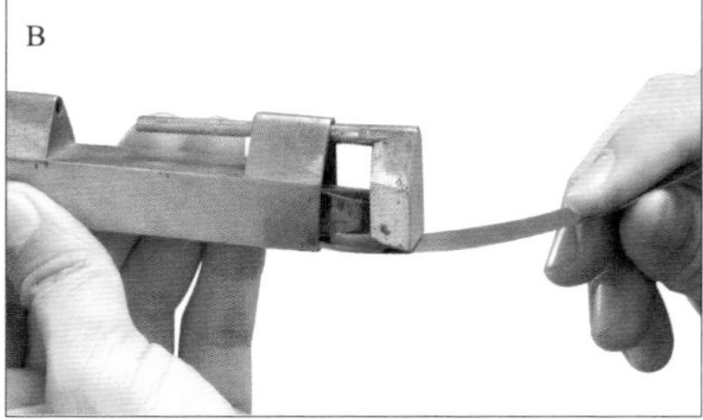

倒拉锁开启步骤

steps of opening a pullback lock

AB

无钥锁 keyless lock

25mm × 78mm × 17mm 65g

参考文献

1. 约瑟夫·N.著，《中国之科学与文明》，第四卷，剑桥大学出版社，1965年。

   Joseph N. , 1965, *Science and Civilization in China*, Vol. IV, Cambridge University Press.

2. 贾杏年，《锁海漫游》，轻工业出版社，北京，1984年。

   Chia, H. N. , 1984, *Roaming the Sea of Locks*, Light Industry Press, Beijing.

3. 颜鸿森，《论中国古代锁具的特点》，第一届中日机械技术史国际会议，北京，1998年10月12日—14日，215—220页。

   Yan, H. S. , 12-14 October, 1998, *On the Characteristics of Ancient Chinese Locks*, Proceedings of the First China-Japan International Conference on History of Mechanical Technology, Beijing, pp. 215-220.

4. 颜鸿森，《古早中国锁具之美》，第一版，中华古机械文教基金会，台南，1999年6月。ISBN 957-97374-9-5。

   Yan, H. S. , June 1999, *The Beauty of Ancient Chinese Locks*, 1st edition, Ancient Chinese Machinery Cultural Foundation, Tainan, ISBN 957-97374-9-5.

5. 颜鸿森，黄馨慧，《古早中国锁具网络博物馆之设计》，第二届台湾机构与机器设计学术研讨会，新竹，1999年12月2日，44—61页。

   Yan, H. S. and Huang, H. H. , 2 December, 1999, *On the Design of a Web-site Museum on Ancient Chinese Locks*, Proceedings of the 2nd Taiwan Conference on Mechanism and Machine Design, Hsing-Chu, pp. 44-61.

6. 颜鸿森，黄馨慧，《中国古代锁具的弹簧配置》，机械与机械原理国际研讨会，卡西诺，87—92 页

   Yan, H. S. and Huang, H. H. , 11-13, May 2000, *On the Spring Configurations of Ancient Chinese Locks*, Proceedings of International Symposium on History of Machines and Mechanisms, Cassino, pp. 87-92.

7. 李如菁，施耐德·R. ，颜鸿森，《锁与钥匙两千年特展专辑》，台湾科学工艺博物馆，高雄，ISBN 957-02-6061-0，2000 年 5 月。

   Li, J. C. , Schneider, R. , and Yan, H. S. , May 2000, *Locks & Keys 2000 Exhibition Catalogue*, National Science and Technology Museum, Kaohsiung, ISBN 957-02-6061-0.

8. 颜鸿森，《古铜锁的收藏与研究》，台湾科学工艺博物馆，高雄，2003 年 3 月。

   Yan, H. S. , March 2003, *On the Collection and Research of Ancient Chinese Locks*, National Science and Technology Museum, Kaohsiung.

关键词

一字锁 "一" letter lock
二开锁 double-open lock
三簧锁 three-spring lock
士字锁 "士" letter lock
上开锁 top-opening lock
上面（钥匙孔）topside (keyhole)
文字组合锁 letter-combination lock
木材（锁）wood (lock)
木锁 wooden lock
方锁 square lock
四开锁 four-open lock
平面钥匙孔 planar keyhole
平面钥匙头（完全包合）
planar key-head (total-enclosure)
平面钥匙头（不完全包合）
planar key-head (partial-enclosure)
平面钥匙头（不包合）
planar key-head (non-enclosure design)
立体钥匙头 spatial key-head
立体钥匙孔 spatial keyhole
多段式开锁 multiple-stage open
百家保锁 hundred-family lock
百家锁 *bai-jia* lock

定向锁 fix-orientation lock
底面（钥匙孔）bottom side (keyhole)
花旗锁 pattern lock
花边锁 pattern-side lock
金（锁）gold (lock)
金属锁 metal lock
红铜（锁）red bronze (lock)
背面（钥匙孔）backside (keyhole)
背开锁 back-opening lock
倒拉锁 pullback lock
栓销制栓锁 pin-tumbler cylinder lock
迷宫锁 labyrinth lock
挂锁 padlock
旋转（开锁）push (open)
组合锁 combination lock
无钥锁 keyless lock
暗门锁 hidden-door lock
景泰蓝（锁）enamel (lock)
单段式开锁 one-stage open
萎蕤琐 solomon's seal lock
黄铜（锁）brass (lock)
玲（锁）ling (lock)
开放式钥匙孔 open keyhole

KEY WORD

寿字锁 "寿" letter lock
端面（钥匙孔）front side (keyhole)
构形（簧片）configuration (springs)
构形（钥匙孔）configuration (keyholes)
管 *guan* (tube)
复合式插入 complex insertion
复合钥匙 compound key
蚀刻 etching
银（锁）silver (lock)
铜锁 bronze (lock)
广锁 broad lock
弹簧片 spring
撑簧锁 prop-open spring lock
虾尾锁 shrimp-tail lock
锒铛锁 *lang-dang* lock
铝（锁）aluminum (lock)
机关锁 puzzle locks
钢（锁）steel (lock)
吓人锁具 scaring lock
隐藏式钥匙孔 hidden keyhole
简单式插入 simple insertion
简单钥匙 simple key
簧片锁 splitting spring lock
转冲锁 rotating-open lock
转轮 rotating wheel
锁 lock
锁栓 sliding bolt
锁体 lock body
镍（锁）nickel (lock)
双开锁 double-open lock
爆仗锁 firecracker lock
绳结 knot
镂刻 engraving
铁（锁）iron (lock)
钥匙 keys
钥匙孔 keyhole
钥匙头 key-head